Underutilized Subtropical Fruits

Sushil Kumar Shukla
Principal Scientist
Central Institute for Subtropical Horticulture
Rehmankhera, Lucknow (UP) India

Sanjeev Kumar
Assistant Director General
U.P. Council of Agricultural Research
Lucknow (UP) India

International Book Distributing Co.
(Publishing Division)

Published by
INTERNATIONAL BOOK DISTRIBUTING CO.
(Publishing Division)
Khushnuma Complex Basement
7, Meerabai Marg (Behind Jawahar Bhawan)
Lucknow 226 001 U.P. (INDIA)
Tel. : 91-522-2209542, 2209543, 2209544, 2209545
Fax : 0522-4045308
E-Mail : ibdco@airtelmail.in

First Edition 2009

Price: Rs. 700/-

ISBN 81-8189-273-9

Composed & Designed at
Panacea Computers
3rd Floor, Agrawal Sabha Bhawan
Subhash Mohal, Sadar Cantt. Lucknow-226 002
Tel. : 0522-2483312, 9335927082, 9452295008
E-mail : prasgupt@rediffmail.com

Printed at
Salasar Imaging Systems
C-7/5, Lawrence Road Industrial Area
Delhi - 110 035
Tel. : 011-27185653, 9810064311

U.P. Council of Agricultural Research
8th Floor, Kisan Mandi Bhawan,
Vibhuti Khand,
Gomti Nagar, Lucknow - 226 010

Prof. Chandrika Prasad
Director General

Foreword

India is one of the 17-mega diverse countries of the world. Though the land area of India is merely 2% of the world, it represents 11% of the known world flora. Besides, there are also wide variations in temperature, humidity, rainfall and soil types. These varied habitats, climatic conditions and soil types offer unique opportunity for commercialization of several underutilized fruit crops which have not been exploited to their full potential so far.

In countries like India the scourge of malnutrition impedes the developmental process. Malnutrition in the form of undernutrition or deficiencies of essential vitamins and minerals continues to cause severe illness or morbidity among millions of peoples. Fruits are considered one of the most important components of human diets as they provide some of the most precious and necessary nutritional components *viz.*, minerals, vitamins, fibres, antioxidants and other micronutrients.

India, a country endowed with diverse climatic conditions, enriched with enormous genetic wealth has immense potential to grow a number of tropical and subtropical fruit crops, which have been neglected so far. The indigenous hardy fruits occurring naturally in one or the other part of the country have always played vital role in nutritional security of rural poor. During last few decades, due to overexploitation of few fruit species/varieties, several less known but nutritionally rich fruit crops have been neglected, resulting in loss of valuable gene pool from the country.

I am glad and appreciate that authors Dr Sushil Kumar Shukla, Senior Scientist (Horticulture), Crop Improvement Division, Central Institute for Subtropical Horticulture, Rehmankhera, Lucknow and Dr Sanjeev Kumar, Assistant Director General, U.P. Council of Agricultural

Research, Lucknow showed their wealth of knowledge and presented several less known fruit crops in a logical sequence by mentioning their importance, the latest cultivars, propagation and production technologies in this publication entitled "Underutilized Subtropical Fruits". All efforts have been made to present the information in an easily understandable form. I am confident that the book will prove useful to all scientists, teachers, students, extension workers, development officers, policy makers and growers alike and thus help promote diversification of agriculture/horticulture, generating employment opportunities for rural youth, improving the economical status of the farmers and ensuring the nutritional security to the rural poor.

2/7/08

(Chandrika Prasad)

Preface

India has emerged as a major stakeholder in the global horticulture scenario accounting for 10.9% of the world fruit production, with the total fruit production of 58.7 million tonnes from 5.5 million hectares. In view of the burgeoning population pressure, it is often not possible to get good piece of fertile land for cultivation of fruit crops, as it is used for other more important field crops, such as foodgrains, oilseeds and pulses.

In the recent years, there has been considerable awareness about the food and nutritional security. A number of underutilized fruits are adequately rich in antioxidants and phytochemicals besides some necessary nutritional components like vitamins, minerals and dietary fibres. It is utmost essential to supplement the diet with fruits, rich in antioxidants to fight and prevent ageing related diseases, obesity and disorders of the body. Hence, more emphasis is being given to the underutilized fruit crops due to their high nutraceutical values in addition to being resistant/tolerant to many biotic and abiotic stresses. The demand for these crops is gradually increasing, resulting in tremendous potential for commercial exploitation of these crops aimed at improving living standards and ensuring nutritional security to the poor and marginal farmers. A good demand for several less known fruits *viz.* aonla, bael, jamun, custard apple, tamarind, carambola, loquat, mulberry, etc. in the fresh and processed form exists in the domestic and the world market. The degraded lands and the areas of water scarcity can profitably be utilized for commercial cultivation of underutilized subtropical fruit crops mentioned in this book.

In the book, attempt has been made by the authors to present the latest information on the cultivars, propagation techniques and production technology of various underutilized subtropical fruit crops *viz.* aonla, bael, carambola, custard apple, wood apple, tamarind, jackfruit, lasora, loquat, barbados cherry, chironji, jamun, phalsa, fig, mulberry, karonda, cape gooseberry, khirni, barhal, amra, star gooseberry and paniyala with suitable photographs for their commercialization. The book will be of immense use to the researchers, students, teachers, policy makers, progressive farmers and all those who are directly or indirectly associated with the development and strengthening of fruit industry.

Authors are highly grateful to Prof Chandrika Prasad, Director General, U. P. Council of Agricultural Research (UPCAR), Lucknow for the sole inspiration to write this book and keen interest in promoting research on underutilized fruits. The authors are also obliged to Dr B. M. C. Reddy, Director, Central Institute for Subtropical Horticulture (CISH), Lucknow; Dr R. P. Shukla, Principal Scientist; Dr Ramesh Chandra, Head, Crop Improvement Division, Dr A. K. Mishra, Dr K. Kumar, Dr G. Pandey, Dr D. Pandey and Dr A. K. Singh, Principal Scientists, CISH for their valuable help and support during the preparation of the manuscript. We are also thankful to Dr R. K. Pathak, ex-Director, CISH, Lucknow; Dr G. C. Tewari, ADG (EPD), Indian Council of Agricultural Research (ICAR), New Delhi, Dr I.S. Singh, ex-Head, Department of Horticulture, NDUAT, Faizabad, Dr P. S. Pathak, ex-Director, IGFRI, Jhansi and Dr Sudhir Kumar Shukla, Senior Scientist (Agronomy), IISR, Lucknow for motivation, moral support and encouragement during the completion of the task. Though this is the first endeavour, constructive criticism will be welcomed by the authors to be included in the next edition.

Lucknow, October 2008

Sushil Kumar Shukla
Sanjeev Kumar

CONTENTS

List of Tables

Chapter 1 Aonla

Botanical Name	:	*Emblica officinalis* Gaertn syn. *Phyllanthus emblica* L.
Regional/ Vernacular Names	:	Amla, aonla, dhatri, dhatriphal, amalaki, nelli, amlakamu, amophal, myrobalan, Indian gooseberry
Family	:	Euphorbiaceae

Aonla fruit is very popular in India due to its therapeutic, medicinal and nutritive value. The hardy nature of tree, high productivity even in marginal soils and suitability of fruit for preparing various cosmetic, herbal and value added products are some of the unique traits which have made aonla popular in the country during last two decades.

Aonla can be adopted successfully for rehabilitation of degraded lands. Introduction of aonla in the range lands of Bundelkhand region, covering parts of Uttar Pradesh and Madhya Pradesh, showed its suitability for hortipasture, agrihorticulture and agrisilvihorticulture land use systems. Aonla has been declared as an ideal fruit crop for various kinds of wastelands, *viz.* moisture stressed areas, eroded lands such as ravines, uplands, riverbeds and areas with undulating topography. It can also be successfully planted on roadsides and strip lands.

Origin and distribution

Aonla originated in tropical south-east Asia, particularly central or southern India. But it is also reported to be the native of Sri Lanka, Malaysia, China. Seedling trees are of common occurrence in mixed dry deciduous forests of India, ascending from sea level to western and eastern Ghats, Aravalli and Vindhyan hills and from north-west to eastern Himalayas.

Aonla is well distributed in India, Sri Lanka, Cuba, Puetrico, Hawaii, Florida, Iran, Iraq, Java, West Indies, Trinidad, Singapore, southern Thailand, Pakistan, Bangladesh, Malaya, China and Panama Canal regions. In India, it is more common in Uttar Pradesh, Madhya Pradesh, Jharkhand, Bihar, Orissa, West Bengal, Karnataka and western Ghats. Moreover, its cultivation is also spreading rapidly in the semiarid regions of Gujarat, Maharashtra, Rajasthan, Andhra Pradesh, Tamil Nadu and parts of Haryana, Punjab and Himachal Pradesh.

Nutritive, medicinal and other uses

The aonla fruit is highly nutritious and one of the richest sources of vitamin C. It contains a fair amount of polyphenols and tannins, which retard the oxidation of vitamin C and render its value as anti-scorbutic in fresh as well as dried condition. Fruits have fair amount of carbohydrates, vitamins and minerals like calcium, iron and phosphorus. Mature aonla fruit contains more vitamin C as compared to developing or immature fruit (Ojha and Pathak, 1993). Aonla fruits are acidic, acrid, cooling, refrigerant, diuretic and laxative (Gopalan and Mohan Ram, 1996). The chemical composition of aonla fruit is presented in Table 1.

Table 1 : Chemical composition of aonla (per 100 g pulp)

Constituent	Content	Constituent	Content
Moisture (%)	77.1 - 82.20	Iron (mg)	1.20
Protein (%)	0.50	Nicotinic acid (mg)	0.20
Fat (%)	0.10	Vitamin C (mg)	200 - 1814
Minerals (%)	0.5 - 0.70	Carotene (mg)	0.01
Fibre (%)	1.9 - 34	Thiamine (mg)	0.03
Carbohydrates (%)	14.10 - 21.89	Riboflavin (mg)	0.05
Calcium (%)	0.012 - 0.050	Niacin (mg)	0.18
Phosphorus (%)	0.020 - 0.026	Tryptophan (mg)	3.00
		Methionine (mg)	2.00
		Lysine (mg)	17.00

Aonla is valued as an anti-scorbutic, diuretic, laxative and antibiotic. One or the other parts of the plant could be used in treating chronic dysentery, bronchitis, diabetes, fever, diarrhoea, jaundice, dyspepsia, cough, etc. It is also being used to treat respiratory and cardiac problems, rheumatism, scurvy and aging ailments. Aonla can fight against diseases because it is rich in vitamin C, tannins and flavanoids, which have antioxidant property. A large number of medicines are being prepared from aonla *viz.*, Chyavanprash, Triphala, Amrit Kalash, Trilax, Amla Plex, Neutrale, Tylophora Plus, Trifla Prash, Trifla Churna, Amalki Ghrit, Brihad Dhatri, Madhumeh Churna, Brihad Dhatri Ghrit,

Bavasir Nashak Mahausidhi, etc. Aonla has tremendous potential in tannin industry due to high tannin content in fruit, leaves and bark. Wild aonla trees could be utilized for this purpose. This can be highly export oriented industry.

Aonla fruits due to high acidity and astringent taste, are rarely used for fresh consumption. The bulk production is to be made into value added products, *viz.* juice, ready to serve (RTS), nectar, squash, syrup, jam, preserve, candy, pickle, sauce, chutney, dehydrated shreds, dehydrated aonla segments *supari*, aonla powder and freeze dried powder. Aonla is used as one of the ingredients in preparation of several cosmetic items, such as hair oil, shampoos, tooth paste, etc.

Soil and climatic requirements

Aonla is a hardy tree mostly cultivated in various kinds of marginal soils. It can withstand slightly acidic to saline/sodic soils in the pH range of 6.0 to 9.5, electrical conductivity (ECe) upto 5 dSm^{-1} and exchangeable sodium per cent (ESP) from 30 to 40. The commercial cultivation has gained popularity in degraded lands like *rakkar*, calcareous, *morum* soils with rocky substratum in Bundelkhand region. In red laterite soils of Maharashtra and Tamil Nadu, aonla has shown better performance as compared to tamarind (Mehta, 2000, Mehta; 2001; Rajput *et al.*, 2000).

Aonla is a subtropical plant which prefers dry subtropical climate. Tree growth is adversely affected by hot winds or frost. However, its cultivation is now quite successful and becoming popular in coastal regions as well as at higher altitude up to 1500 m above sea level (ASL), but aonla plants are susceptible to frost during winter in north India. In Rajasthan, Gujarat and Maharashtra, extremely high temperature coupled with desiccating winds during summer is very common and sometimes it causes excessive blossom and pin head drop. Temperature above 45°C and low humidity even for few days may show injurious influence and adversely affect the bearing. Low temperature around 7 to 10°C is conducive for floral bud initiation, however warm temperature and humid atmosphere is conducive for fruit growth and development. Dry spell during rainy season results in heavy dropping of fruitlets and prolonged fruit dormancy.

Cultivars

Banarasi, Bansi Red, Francis *(Hathijhool)* and Chakaiya are the old known aonla cultivars. All these cultivars have their own merits and demerits. Banarasi is a shy bearing cultivar and prone to heavy fruit drop; Francis, though prolific bearer, has serious incidence of fruit necrosis. Chakaiya, the most popular cultivar has comparatively smaller sized fibrous fruits with a tendency to bear in alternate years (Pathak *et al.*, 2003). The improved cultivars of aonla as described by Pathak *et al.* (1993) and Chundawat (1990) are given below with their chief characteristics:

Kanchan (NA-4)

It is a seedling selection from Chakaiya. The cultivar is heavy and regular bearer. Fruits are medium sized with high fibre content, preferred by industry for pulp extraction and preparation of value added products. It has mid season maturity (mid November to mid December) and has adopted well in the arid and semi-arid regions of Gujarat and Maharashtra.

Krishna (NA-5)

Krishna is a seedling selection from Banarasi. It is an early maturing genotype, which matures during mid October to mid November. The fruits are large, triangular, conical, smooth skinned, whitish green to apricot yellow with red spot on exposed portion. Flesh is pinkish green, less fibrous and highly astringent with moderate keeping quality. It is an ideal variety for preparing candy and juice.

Narendra Aonla-6 (NA-6)

This is a selection from Chakaiya which has mid season maturity (mid November to mid December) and heavy bearing. Fruits are most attractive and shining, medium to large size, flattened and very low in fibre content. This is an excellent variety for making preserve, candy and jam.

Narendra Aonla-7 (NA-7)

It is a seedling selection from Francis which is precocious, prolific and regular bearer. Incidence of necrosis has not been observed in this cultivar. It has mid season maturity. Fruits are medium to large in size with conical apex. Fibre content is slightly higher than NA-6. This variety has adopted well in the states of Rajasthan, Bihar, Madhya Pradesh, Andhra Pradesh and Tamil Nadu. Staking of the fruited branches is desirable in this cultivar as tree branches often give way due to heavy fruit load. Fruits are suitable for making Chyavanprash, chutney, pickle, jam and squash.

Narendra Aonla-10 (NA-10)

This is a chance seedling selected from cultivar Banarasi. It bears profusely and fruits are attractive, medium to large in size and flattened round in shape. Skin is rough, yellowish green with pink tinge. Flesh is whitish green, fibre content is higher and phenolic content is lower. This is an ideal cultivar for commercial growing because of its higher productivity, early maturity, suitability for dehydration and pickle making. However, it has a tendency to bear in alternate years.

BSR-1

This is also a seedling selection from Tamil Nadu. It is a self-fruitful cultivar which is prolific bearer and gives crop round the year. Fruits are small (27 g) but with high vitamin C content (620 mg per 100 g), fibrous and low moisture content. The cultivar has bright prospects for use in preparation of ayurvedic medicines/tonics.

Propagation

Aonla has long been raised by seed. Though seed propagation is easy and cheaper, it does not ensure true to type plants due to cross pollination. Besides, such trees take long time to come into bearing. Aonla was earlier propagated through inarching, but owing to upright growth habit, availability of scion shoots has been the limiting factor. In the recent years, techniques of patch budding, soft wood grafting, cleft/wedge grafting in polythene bags raised rootstocks have been standardized and are being advocated for commercial multiplication in aonla growing regions of the country. Cleft/wedge grafting can be done almost throughout the year using propagation structures like net house, polyhouse, mist chamber, etc., while patch budding gives very good success during June to September in both field and polythene bags.

Six month to one year old, vigorous and pencil thick aonla seedlings are used as rootstock. For seed extraction, mature fruits collected from seedling aonla trees during January-February are dried in the sun and seeds are extracted using light pressure. Sowing of seed can be done either during February-March or June-July in raised beds or polythene bags. When raised in seed beds, seedlings should be shifted to field or perforated polythene bags at 2 to 4 leaf stage or 1 to 1½ month after sowing. Perforated nursery polythene bags of 25 cm height and 10 to 15 cm diameter should be filled with soil and well rotten farm yard manure (FYM) in equal proportions. In case of clay or heavy textured soil, equal quantities of soil, FYM and sand should be mixed to prepare the soil mixture. The seedlings are ready for grafting/budding in 6 to 12 months. Seedlings planted in field attain graftable thickness earlier than the ones planted in polythene bags.

The method of bench grafting has also been standardized for aonla during spring season. This method involves cleft grafting on bare rooted, pencil thick aonla seedlings followed by planting in polythene bags, covering the scion and grafted portion with polythene tube, keeping them in net house or partially shaded place followed by watering. It gives very good success during spring season (February-March). Use of polythene tube for covering the scion effectively prevents the transpiration losses and drying of scion leading to better success. Bench grafted three months old aonla plants are ready for sale.

Planting and after-care

Grafted or budded plants of aonla are planted 8 to 10 m apart in square system of planting during July-August or February. For high density planting of aonla, rows are spaced 8 m apart, while plant to plant distance is reduced to 4 m accommodating 312 plants per ha. In wastelands, pits of one cubic meter size are dug two months prior to planting. The pit soil is divided into two equal parts, *i.e.*, 50 per cent top and 50 per cent bottom soil. In the top soil, 3 to 4 basketfuls (25 to 40 kg) of well rotten farm yard manure (FYM) and 1 kg neem cake or 500 g bone-meal is mixed. The pit is filled with the mixture 10 to 15 cm raised from the ground level. In sodic soils, as per gypsum requirement (GR) value, 5 to 8 kg gypsum or pyrite or "*Usar Tor Khad*" along with 20 kg sand is also incorporated in the filling mixture. Filled pits are irrigated thoroughly, if there is no rain so that the soil gets properly set after filling. In sodic soils, planting should be done 20 to 25 days after filling to complete the soil reaction with gypsum or pyrite. The budded/grafted plants with earth ball should be planted in the centre of the pits. The soil of the pit should be properly pressed around the earth ball. Light irrigation should be given just after planting. If grafted plants are in polythene bags (polybags), the bag should be removed carefully before planting. The polythene strip, if at all present around the bud/graft union, should also be removed at the time of planting. Self-incompatibility in aonla is a common problem among various cultivars, hence, two varieties should be planted in alternate rows for higher productivity.

In degraded, arid, semi arid lands, ravines, *kandi* and *ghar* (foothills of

Himalayas) areas, *in-situ* orchard establishment has given encouraging results. In this technique, sowing of 2 to 3 seeds in well prepared pits is done. Care should be taken to encourage vigorous seedling growth. In saline/sodic soils, seed germination is adversely affected. Under such conditions, seedlings are raised in polybags/containers. After gaining appropriate height, these seedlings are transferred and following proper establishment, these are budded with the scion shoots from improved aonla cultivars. Since, scion of aonla can safely be stored/ transported in moss grass/moist gunny bag for 5 to 7 days, this practice is helpful in establishment of aonla orchards in distant places (Pathak *et al.*, 2003). This technique has also given very good results under Jhansi (Uttar Pradesh) conditions when compared with the traditional technique (Shukla and Solanki, 2000).

The systematic work lacks on nutritional requirements of aonla trees. The fertilizer doses depend on soil fertility, age of the plant and production. Based on experience, recommendations of manure and fertilizers are given in Table 2.

Full dose of well rotten FYM, half dose of nitrogen and full doses of phosphorus and potassium should be applied during January-February

Table 2 : Manure and fertilizer requirements of aonla

Age of tree (years)	FYM (kg)	Nitrogen (g)	Phosphorus (g)	Potassium (g)
1	5	100	50	100
2	10	200	100	200
3	15	300	150	300
4	20	400	200	400
5	25	500	250	500
6	30	600	300	600
7	35	700	350	700
8	40	800	400	800
9	45	900	450	900
10 or above	50	1000	500	1000

before flowering. Remaining half dose of nitrogen is given during July-August. The manure and fertilizers are mixed in the soil under the tree canopy leaving one meter around tree trunk. Light irrigation is given immediately after application of fertilizers. In sodic soils, 100 g each of borax, zinc sulphate and copper sulphate should be applied along with fertilizers in accordance with tree age and vigour. Besides this, spray of micronutrients, *viz.* boron (0.4%), zinc (0.4%), copper (0.4%) along with lime is also recommended. It is helpful in reducing fruit drop and improving fruit quality.

Aonla is a hardy tree. Established aonla orchards, in general, do not require additional irrigation except normal rains. However, irrigation at an interval of 15 to 20 days is required during winter (October to January) and summer season (April to June), particularly during initial years of orchard establishment. It protects the trees against frost and also controls fruit drop during winter. Brackish water should not be used for irrigation. In bearing plantations, first irrigation should be given just after fertilizer application (January-February). Irrigation should be avoided during flowering (mid March-mid April) period. Mulching with organic wastes in tree basin has been found very effective for establishment of aonla orchards, suppressing weed growth and conserving soil moisture.

Aonla plants should be encouraged to develop into a tree of medium height. Young plants are allowed to grow straight up to 0.75 to 1 m from the ground and then allowed to branch. Three to four branches with wide crotch angle are encouraged in different directions in a well distributed manner during initial years. Staking of heavily fruited branches is done to avoid damage to the tree branches. Regular pruning of a bearing aonla tree is not required. Annual shedding of determinate shoots encourages new growth. However, dead, infested, broken, diseased, weak, criss-cross branches and suckers developing from rootstock should be removed regularly.

Old seedling trees or other inferior quality aonla trees can be top worked or rejuvenated and easily changed into superior types by top working. These plants are headed back 3 to 4 m above the ground level during December-January, leaving 4 to 6 main cut limbs in all the directions. The cut ends of the trees are then pasted with the wet soil or fresh cow dung. Only four to six shoots, developing in outer directions on main limbs, are allowed to develop. During June-July, patch budding with superior scion cultivars is done on these shoots. After sprouting, the top portion of the budded shoots is removed. Care is taken to remove rootstock suckers regularly and manage insect pest problems.

Harvesting and yield

Aonla tree behaves as an evergreen plant in the tropics, but under subtropical conditions of north India, it begins to shed its determinate shoots along with leaves from November onwards and the indeterminate shoots are devoid of foliage by February-March. New determinate shoots begin to appear by the end of February and continue to emerge till first week of April (Bajpai, 1965; Ram, 1971). The flower buds appear on these newly emerged shoots. In north India, flowering takes place from March to May, while in south India, it blooms in June-July and again in February-March. Male flowers appear in clusters in leaf axils on lower parts of the determinate shoots followed by female flowers on same shoots. However, some of the shoots bear only male flowers and those appearing late in the season do not flower. Aonla is a cross pollinated fruit plant and wind, honey bees and gravity play important role in effective pollination. The self incompatibility problem was confirmed in aonla (Balasubramaniam and Arulmozhiyan, 2003, Singh *et al.*, 1998). Therefore, there is a need for provision of proper pollinizer in order to improve fruit set and retention. NA-6 and Krishna were found suitable pollinizers for NA-7.

Aonla fruits are harvested at full maturity. The maturity can be judged by one of the following criteria *i.e.*, specific gravity (1.07 to 1.10), TSS/acid ratio (5 to 6), change in fruit colour (green to shining whitish green or yellowish green) and change in seed colour (from creamy white to brown). Apart from these indices, maturity of the commercial cultivars has been worked out, *e.g.* Banarasi and Krishna mature 17 to 18 weeks after fruit set, while Kanchan and Francis take about 20 weeks. Chakaiya matures 23 weeks after fruit set. The aonla fruits are usually hand picked climbing up the tree. The fruits should be harvested in the morning hours and be collected in plastic crates. The fruits should not be allowed to drop on the ground as the injured fruits cause spoilage to other healthy fruits during packaging and storage.

A budded/grafted aonla tree starts bearing 3 to 4 years after planting, whereas a seedling tree may take 6 to 8 years. Vegetatively propagated plants attain full bearing after 10 to 12 years. The yield varies greatly in different aonla cultivars. Banarasi is a shy bearer, Francis and NA-6 are average bearer, while Kanchan and NA-7 are prolific bearer. Number of factors *viz.*, genetic potential, environment and management contribute to the production. A full grown aonla tree produces 100 to 300 kg fruits per tree, thus gives the fruit yield of 15 to 20 tonnes per ha.

The fruits of aonla cultivars Chakaiya and Francis can be stored up to 9 days and Banarasi up to 6 days at ambient temperature. Under zero-energy cool chamber, fruits of Chakaiya cultivar can be stored up to 16 days with minimum decay and quality loss (Nath *et al.*, 1992). Studies conducted on aonla cultivar Chakaiya showed that the aonla fruits can be stored at least 3 months in 15% salt solution with minimum quality (appearance and vitamin C content) loss (Kumar *et al.*, 1992).

References

Bajpai, P.N. (1965) *Agra Univ. J. Res. (Sci)* **14**: 167-186.

Balasubramaniam, S. and Arulmozhiyan, R. (2003) Paper presented in Nat. Sem. on Production and Utilization of Aonla, organized by All India Aonla Grower's Association, Salem, Tamil Nadu, India.

Chundawat, B.S. (1990) Arid Fruit Culture, Oxford and IBH Publishing Co. Pvt. Ltd., New Delhi. pp. 110-118.

Gopalan, I. and Mohanram, M. (1996) Fruit, National Institute of Nutrition. ICMR, Hyderabad, pp. 3-4.

Kumar, S., Nath, V., Singh, I.S. and Nainwal, N.C. (1992) *Prog. Hort.* **24** (1-2): 83-86.

Mehta, S.S. (2000) Brainstorming session on the exploitation of the potential of arid fruit in Tamil Nadu, organised by Tamil Nadu Fruit Growers Association, Dindigul, Tamil Nadu. Nov. 2000, pp. 51-55.

Mehta, S.S. (2001) A case study on development of aonla in Tamilnadu. Proc. Nat. Hort. Conf. held at New Delhi, Nov. 2001 (Eds. H.P. Singh, J.P. Negi, and J.C. Samuel). pp. 160-163.

Nath, V., Singh, I.S. and Kumar, S. (1992) *Narendra Deva J. Agric. Res.* **7**(1): 121-126.

Ojha, C.M. and Pathak, R.K. (1993) *Prog. Hort.*, **25**(1-2): 42-46.

Pandey, D., Pathak, R. K. and Tandon, D.K. (2005) *Indian Horticulture*, **50**(1):20.

Pathak, R.K., Pandey, D., Haseeb, M. and Tandon, D.K. (2003) The Aonla, CISH, Lucknow, India.

Pathak, R.K., Srivastava, A.K., Dwivedi, R. and Singh, H.K. (1993) Aonla Descriptor Res. Bull., Directorate of Res., NDUAT, Faizabad.

Rajput V., Dhake, A.V. and Balasubramanyam, V.R. (2000) *Plant Hortitech.*, **1**(5):49-52.

Ram, S. (1971) Ph.D. Thesis, Kanpur University, Kanpur.

Shukla, S.K. and Solanki, K.R. (2000) *Indian Hort.*, **44**(4): 12-13.

Singh, H.K., Srivastava, A.K. and Dwivedi, R. (1998) *Ind. J. Agric. Sci.*, **68**(4) : 204-205.

Chapter 2 Bael

Botanical Name	:	*Aegle marmelos* Correa
Regional/ Vernacular Names	:	Bael, bilva, shul, bel, bil, maredu, vilvam, bilpatre, Bengal quince, Indian quince, golden apple, holy fruit, stone apple, belwa, sriphal
Family	:	Rutaceae

Bael is one of the oldest and useful trees of India. All the parts of bael tree have their own importance. It is medicinally and economically important fruit crop with immense potential for extending its cultivation in arid and semi-arid degraded lands. Bael fruit is available in almost all states of India but most abundantly in Uttar Pradesh, Madhya Pradesh, Bihar, Jharkhand, West Bengal and Orissa.

Origin and distribution

Bael is indigenous to India and known since prehistoric times. The tree grows wild in dry forests, on hills and plains of central and southern India, Myanmar, Pakistan and Bangladesh also in mixed deciduous and dry forests of former French Indochina. It is cultivated throughout India and main growing states are Uttar Pradesh, Bihar, Jharkhand, Madhya Pradesh, West Bengal, Orissa, Chattisgarh and Rajasthan. Because of its status of sacred tree, it is also grown in Sri Lanka and northern Malaya, the drier parts of Java and to a limited extent in northern Luzon in the Philippine Island. Systematic orcharding of bael is extremely limited, however it is generally grown in parks, temple gardens, roadside, backyards of house, etc.

There is no organized orcharding of bael, hence exact data on area and production are not available. Some progressive orchardists have also started planting improved bael cultivars in form of small orchard or as boundary plantation. Most of the bael fruits available in the market are harvested from the trees growing naturally in the forest area or from the seedling plantations.

Nutritive, medicinal and other uses

The Bael fruit is one of the most nutritious fruits. It contains 61.5 g water, 1.8 g protein, 0.39 g fat, 1.7 g minerals, 31.8 g carbohydrates, 55 mg carotene, 0.13 mg thiamine, 1.19 mg riboflavin, 1.1 mg niacin and 8 mg vitamin C per 100 gm of edible portion (Gopalan *et al.*, 1971). It is a rich source of riboflavin among fruits. Marmelosin is, most probably, the therapeutically active principle of bael fruit. It has been isolated as a colourless crystaline compound (Dixit and Dutt, 1932). The astringent taste of bael fruit is due to presence of phenolic compounds. The medicinal properties of bael have been dealt with in Charaka Samhita (Aiyar, 1956). Bael fruit has curative, pesticidal and nutritive properties. The leaves of bael are used for worship of Lord Shiva. Green mature bael fruits are used for preparing *murabba* (preserve) which is generally taken for stomach ailments. Bael fruits have immense potential for processing into products such as squash, jam, toffee, candy and syrups for internal and foreign market.

Fruits have mucilaginous substance. Gum is also extracted from seed locules. The gum is used for adhesive, water proofing and oil emulsion coating, whereas the wood is used for making small agricultural implements. The leaves are valued as fodder for sheep and goats. The leaves contain 0.6% of essential oil, mostly limonene. Dry leaf powder is used as a protectant against storage pests of paddy. The seed contained 62% protein (60% water insoluble and 2% water soluble), 32% oil, 3% carbohydrates and 3% ash (Banerjee and Maiti, 1980).

Two important homoeopathic medicines namely *Aegle marmelos* and *Aeglefolia* are made from bael fruit pulp and leaves, respectively. The *sharbat* made from bael pulp is taken for mild laxative, tonic and digestive effects. Decoction of unripe fruits mixed with fennel and ginger is prescribed in cases of haemorrhoids. Marmelosin extracted from pulp is used as laxative and diuretic and, in large doses, lowers the rate of respiration, depress heart action and cause sleepiness. The tender fruits are sliced horizontally, sun dried and sold in market and also exported to Malaya and Europe. Oil extracted from seed is prescribed as purgative. It contains 15.6% palmitic acid, 8.3% stearic acid, 28.7% linoleic acid and 7.6% linolenic acid. The leaf juice mixed with honey is given to allay catarrh fever and, if mixed with black pepper, relieves jaundice and constipation accompanied by oedema. A hot poultice of the leaves is considered effective for ophthalmic and various inflammations and also febrile delirium and acute bronchitis. Decoctions of the root are advocated to relieve palpitation of the heart, indigestion and bowel inflammations and also to overcome vomiting.

The bael fruit is most commonly used for making *sharbat* as it is cooling, refreshing, highly nutritious during hot summers. Most of the bael produce received in major fruit markets of Uttar Pradesh, Bihar, Madhya Pradesh and Jharkhand is used for *sharbat* making. The *murabba* (preserve) made from mature fruits is the most common processed bael product. A beverage is also made by combining bael pulp with that of tamarind. Besides these, a number of value added products, namely *i.e.* squash, fruit slab, toffee, powder, jam, etc. are prepared and the recipes have been suggested by Singh and Roy (1984). Ram and Singh (2003) screened different bael varieties for processing and found NB-9 good for processing.

Soil and climatic requirements

Bael is a hardy fruit tree which has wide adaptability to adverse soil and climatic conditions. It can thrive well in saline, alkaline, rocky, stony, degraded soils and ravines. However, a well drained sandy loam soil is ideal. Heavy soils or areas with high water table should be avoided for bael. It can tolerate sodicity up to 30 ESP, salinity up to 9 dSm^{-1} and a pH range of 6.0 to 8.5 (Singh, 2001). Bael plants can grow easily in waste lands and in sandy soils of low fertility and poor water holding capacity even in Thar desert.

The bael trees grow successfully and prefer subtropical climate where summer is hot and dry and winter is mild. It can be grown up to an elevation of 1200 m above sea level (ASL) and is not injured by a temperature as low as -7^0C and as high as 48°C. Plants tolerate high temperature and drought during summer months by shedding leaves.

Cultivars

Old cultivars of bael were known after the names of the places, where they were dominant. Six cultivars of bael were described by Singh (1961) from Uttar Pradesh and Mirzapuri was considered to be the best followed by Darogaji, Ojha, Rampuri, Azamati and Khamaria. Teaotia *et al.* (1963) also listed five cultivars from Uttar Pradesh and found that Kagzi Gonda was the best producing large fruits (1412.82 g per fruit) with thin rind (0.15 cm) and soft yellow pulp of excellent flavour. Jauhari and Singh (1971) surveyed the important fruit growing areas of Uttar Pradesh and Bihar and found that among the cultivars studied, Kagzi Etawah, Sewan Large, Mirzapuri and Deoria Large were excellent in taste and other qualities. Mazumdar (1975) studied five cultivars of bael from West Bengal and found the spherical flattened ones were usually the best on the basis of fruit and chemical composition. Singh and Kumar (1991) studied thirteen germplasms of bael from eastern Uttar Pradesh and found the spherical one was superior on the basis of physico-chemical composition of fruits. Improved bael cultivars developed by N.D. University of Agriculture and Technology (NDUAT) Faizabad; G.B. Pant University of Agriculture and Technology (GBPUAT), Pantnagar and Central Institute for Subtropical Horticulture (CISH), Lucknow are gaining popularity. Chief characteristics of these cultivars as described by Singh and Ali (1992); Mishra (2007) and Pathak *et al.* (2003) are given below:

Narendra Bael-5 (NB-5)

The plants are dwarf and spreading. They are precocious and prolific in bearing. The average fruit yield of eight-year old plant is 56 kg. The fruits are medium in size (1.25 kg per fruit), round, with smooth surface and thin rind (2.5 mm), yellow at maturity, low in mucilage, moderately fibrous and attractive yellow pulp with low seed content. Excellent in taste and flavour, the fruits have 38 per cent total soluble solids, slightly acidic in taste and contain 25 mg ascorbic acid per 100 g pulp.

Narendra Bael-9 (NB-9)

The plants are semi-vigorous and spreading. The variety is precocious and prolific bearer. The average fruit yield of eight-year old tree is 65 kg. Fruits are medium to large in size (1.5 kg per fruit), roundish-oblong, with smooth surface and thin rind (3 mm), light yellow at maturity, average in mucilage, moderately fibrous and golden-yellow pulp with low seed content. The fruits are good in taste containing 40 per cent total soluble solids, slightly acidic and 28 mg ascorbic acid per 100 g pulp.

Pant Aparna

Trees are dwarf with drooping sparse foliage, almost thornless, precocious and heavy-bearer. The leaves are large, dark green and pearshaped. Fruit is globose in shape, with average weight of 0.6 to 0.8 kg. Fruit pulp is yellow and rind is thin. Fruit has 46% flesh, 34% TSS, 0.8% acidity and 15 mg ascorbic acid per 100 g of flesh. Mucilage, seed and fibre are low. Mucilage and seeds are enclosed in separate segments. Flavour and taste are very good.

Pant Shivani

It is a mid-season variety. Trees are tall, vigorous, dense, upright growing, precocious and heavy bearer. Fruit shape is ovoid to oblong. Average fruit weight is 2 kg. Colour of fruit is lemon-yellow and good storage quality. Rind is medium-thin, pulp is lemon-yellow with pleasant flavour and mucilage, seeds and fibres are low to medium. It tastes very good. It has 69% flesh, 34% TSS, 1.31% acidity and 15 mg ascorbic acid per 100 g of flesh.

Pant Sujata

It is an early mid-season variety but has problem of fruit splitting. Trees are medium-dwarf with drooping and spreading foliage, dense, precocious and heavy-bearer. Thorns are stout and bigger. Fruit is globose, depressed on both ends, with average weight of 1.14 kg. Fruit and pulp are light yellow. Storage quality is better, rind is thin, and seeds, mucilage and fibres are low. Its flavour is pleasant and taste is very good. Fruit contains 72% flesh, 30% TSS, 0.75% acidity and 18 mg ascorbic acid per l00 g of flesh.

Pant Urvashi

It is a mid-season variety. Trees are tall, vigorous, dense, upright growing, precocious and heavy bearer. Fruit is ovoid-oblong with average weight of 1.6 kg. Fruit is lemon-yellow, rind is medium thin and pulp is yellow. Fruit has 64% flesh with pleasant flavour. Seed and mucilage contents are medium, fibre content low, TSS 33%, acidity 1.15% and ascorbic acid 12.6 mg per l00 g of flesh.

CISH B-1

Plants are moderate in height and less spreading. Fruits are oval to oblong having size of 15-17 X 39-41 cm spread, good in taste (35 to 40.5^0Brix TSS). The average weight of fruit is 0.8 to 1.2 kg. Fruit rind is thin (0.10 to 0.12 cm) and weight of fruit shell is 125 g per fruit. Fibre and seed contents low and average yield per tree is 50 to 60 kg.

CISH B-2

Plants are semi vigorous and spreading. Fruits are bigger in size, round or oblong in shape. The weight of fruit is 1.8 to 2.7 kg. Fruits are good in taste with 37 to 41^0Brix TSS. The fruit has thin rind, scanty fibre and few seeds. The average yield per tree is 40 to 50 kg.

Propagation

Earlier, when there were no standardized vegetative propagation methods for bael, it was multiplied by seeds. Plants of seedling origin exhibit a lot of variability in plant and fruit characteristics. Now, seedlings are commonly used as rootstock for budding/grafting of improved cultivars. For raising seedlings, the freshly extracted seeds from healthy fruits should be sown immediately during monsoon season. Since bael seeds are recalcitrant, they cannot be stored for longer periods under normal storage conditions. To prevent the losses due to the 'damping off' in nursery seed treatment with Thiram (1:400) is essential.

Seeds are sown either in raised beds or polythene bags. As bael fruit matures during March-April, freshly extracted seeds should be sown during June-July. Since seed size is small, seeds should not be placed very deep in soil. Immediately after sowing, light irrigation and mulching promote germination. When seedlings attain 10 to 15 cm height in nursery bed, these should be dug up along with earth ball and planted in field or polybags for grafting after attaining suitable thickness. The shifting of bael seedlings, at intervals, is necessary to check the growth of tap root system and encourage fibrous/ adventitious roots.

Bael can also be propagated successfully by budding on 1 year old rootstocks. Success per cent of budding varies with season. Various workers have reported that period from mid May to July is the most congenial for getting higher percentage of success. Moti *et al.* (1976) reported superiority of patch budding over shield method. June was found to be the best period of budding. Singh *et al.* (1976) propagated bael plants by patch budding, T budding and chip budding at monthly intervals from July to October and from March to June. Bud take was 100% with patch budding in June and July. Scion growth assessed 90 days later, was greatest with July budding. Kumar *et al.* (1994) found May-June as the best time and patch budding as the best method for bael propagation under Faizabad conditions. Singh and Kumar (1991) reported that patch budding performed during the month of June gave maximum success (90%) than the other methods (shield and ring budding) and timings (July and August). Patch budding on 1 to 2 year old rootstocks in June, using one month old scion gives 100% success (Mishra, 2007). Old and uneconomic bael trees can also be top worked using this method. The trees are headed back 1 to 1.5 m above the ground level during March and new shoots emerging from the stump are patch budded with suitable scion variety in the following June-July.

Soft wood grafting using cleft method gave 95% success in bael (Mishra, 2007; Ram and Pathak, 2006) and it is being advocated for commercial bael propagation in north India. This technique reduces the time period considerably for preparing new grafts as compared to budding. With the use of mist chamber, polyhouse/shade net house, the bael can be propagated throughout the year using this technique.

Planting and after-care

Grafted or budded bael plants are planted at the distance of 8 x 8 m during July-August or February-March. Pits of one cubic meter size are dug two months prior to planting. The hard pan/rock, if present in the soil must be broken. The pits are filled with top soil and well rotten farm yard manure mixture in 1:1 ratio. It is advisable to mix 250 g single super phosphate and 100 g neem cake in soil while filling the pit. In sodic soil, 5 to 8 kg gypsum along with 20 kg sand is also incorporated as per gypsum requirement. The pits are irrigated thoroughly after filling, if there are no rains to settle down the soil in the filled pit. After the soil becomes workable, the plants are planted in the centre of the pits. After planting, irrigation should be given immediately and repeated at weekly intervals till the plants get established. In salt-affected soil, planting should be done 20 to 25 days after filling to complete the soil reaction with gypsum or pyrite. In marginal and degraded lands with paucity of irrigation water, orchard establishment should be encouraged *in-situ*. As per layout plan, bael seeds are sown in the pit after filling with mixture in June-July. Polythene raised seedlings can also be planted in the field. Patch budding with suitable variety is performed in the following year in May-June.

For optimum growth, yield and quality production of bael, manure and fertilizer application should be done judiciously. A dose of 5 kg well rotten FYM, 100 g nitrogen, 50 g phosphorus and 50 g potassium should be applied to one year old plant. This dose should be increased every year in the same proportion upto the age of 10 years. Thereafter, a constant dose of 1000 g nitrogen, 500 g phosphorus and 500 g potassium along with 50 kg FYM should be applied. In sodic soils, plants generally show zinc deficiency symptoms, for which 250 g zinc sulphate per plant along with fertilizer application is also recommended. Besides this, foliar application of 0.5% zinc sulphate in

July, October and December is also recommended. Full doses of manures and fertilizers should be applied in the month of June-July. The manure and fertilizers should be spread under the canopy of the tree and incorporated well in the soil. In bael orchards, where fruit cracking is a problem, 300 g borax per tree should also be applied along with manure and fertilizers. After manure and fertilizer application light irrigation is essential.

Plants should be watered as and when needed. Basin system ensuring uniform distribution of water should be used for irrigation of young plants. Light irrigations should be given to the young plants frequently. In areas of water scarcity, pitcher irrigation is successful for bael orchard establishment. In bearing orchards, irrigations at 15 to 20 day intervals in summer and at monthly intervals in winter are recommended. Tree basins should be cleaned from time to time by regular weeding. Mulching with organic wastes has been found very effective for establishing bael orchards in saline, sodic and eroded lands. Paddy straw, sugarcane trash, FYM, dry leaves can be used for the purpose. It helps in improving organic matter content, conserving soil moisture and restricting weed growth.

Young plants are trained with the help of stakes, so that they grow erect. In order to provide good framework, it is essential not to allow lateral branches upto 75 cm from the ground. Afterwards, 4 to 6 branches arising from different directions should be allowed. The tree should be trained in modified leader system during initial 4 to 5 years. Generally, bael does not require pruning. However, dead, dense and criss-cross branches, weak and broken twigs are pruned off once a year after harvest. During the early years of young plantation, the interspace should be utilized for growing suitable intercrops. Leguminous crops like pea, cowpea, moong, urd and guar; and vegetables like brinjal, tomato, spinach, coriander, chilli, garlic, etc. can be grown preferably as intercrops. The crops having high water requirement should be avoided, since bael is susceptible to water logging. In salt affected or marginal soils, it is advisable to grow green manuring crops like dhaincha (*Sesbania sp.*) for few years to improve physico-chemical properties of the soil.

Harvesting and yield

Flowering starts with the commencement of new growth in May-June. Fruit setting takes place by the end of May and continues up to July. Fruiting occurs on one year old shoots. Fruits are ready for harvest in

April-May when the fruit shell changes colour from deep green to yellowish green. The fruit should be picked individually, so that it does not fall on the ground to avoid cracking of the shell or any physical impact which may lead to spoilage during storage. The stem end of the fruit is prone to infection, therefore, the fruits should be harvested along with 1 to 2 cm of stem end. The stalk gets automatically separated from the fruit when the fruit is ripe, thus it is also an indication of ripening. Bearing in budded/grafted plants starts 3 to 4 years after planting, whereas seedling trees may take 7 to 8 years. The number of fruits per tree increases with the age of the tree. A 10 to 15 year old fully grown up tree can yield upto 150 to 250 fruits. A full grown bael tree of 30 to 35 years may produce 1000 to 1500 fruits.

Bael fruits can also be harvested as early as in January and ripened artificially using ethrel (Mishra, 2007). Thus, harvesting time can be advanced by 2 to 3 months. For ripening, fruits are treated with 1000 to 1500 ppm ethrel and stored at 30^0C. It takes 18 to 24 days to ripen the fruits artificially.

References

Aiyer, A.K.Y.N. (1956) The antiquity of some field and forest flora of India, Bangalore, Printing and Publising Co. Ltd., Bangalore.

Banerjee, N. and Maiti, M. (1980) *J. Indian Chem. Soc.*, **57**(9):945-946.

Dixit, B.B.L. and Dutt, S. (1932) *J. Indian Chem. Soc.*, **9**: 271-279.

Gopalan, C., Rama Sastri, B.V. and Bala Subramanian, S.C. (1971) Nutritive Value of Indian Foods, NIN, ICMR, Hyderabad.

Jauhari, O.S. and Singh, R.D. (1971) *Indian Hort.*, **16**:9-10

Kumar, D., Pathak, R.K. and Ali, W. (1994) *Indian J. Hort.*, **51**(2): 150-153.

Mazumdar, B.C. (1975) *Indian Agric.*, **19**(3): 295-298.

Mishra, K.K. (2007) *Indian Hort.*, **52** (3) : 13-15.

Moti, L. D. and Chaturvedi, O. P. (1976) *Punjab Hort. J.*, **16**: 57-59.

Pathak, R. K., Pandey, D. and Pramanick, P.K. (2003) *ICAR News*, **9**(1): 16.

Ram, D. and Singh, I.S. (2003) *Prog. Hort.*, **35** (2): 199-201.

Ram, R.A. and Pathak, R.K. (2006) *Indian Hort.*, **51**(4) : 10-11.

Singh, I.S. (2001) Bael. In: Handbook of Horticulture, (Ed. K.L. Chadha), ICAR, New Delhi, pp. 140-142.

Singh, I.S. and Ali, W. (1992) *Indian Hort.*, **37**(3) : 35.

Singh, I.S. and Kumar, S. (1991) *Nw Agriculturist*, **2**(2): 191-194

Singh, L.B. (1961) Annual Report. Hort. Res. Inst., Saharanpur : 111-119.

Singh, R.N. and Roy, S.K. (1984) The Bael, ICAR, New Delhi, p. 25.

Singh, U.R., Pandey, I.C., Upadhyay, N.P. and Prasad, R.S. (1976) *Punjab Hort. J.*, **16:** 57-59.

Teaotia, S.S., Maurya, V.N. and Agnihotri, B.N. (1963) *Indian J. Hort.*, **20** : 210-214.

Chapter 3 Carambola

Botanical Name	:	*Averrhoa carambola* L.
Regional/Vernacular Names	:	Kamarakh, star fruit, five corner fruit
Family	:	Oxalidaceae

Carambola is an important fruit of warm tropical and subtropical areas of the world. The tree is medium sized, attractive and evergreen growing to about 10 m height. Organized orcharding of this fruit in India is rare but it is often planted as wind break tree and commonly found in several gardens of tropical and subtropical parts of our country. The fruit is emerging as an interesting addition to the table fruit, not only as a fresh fruit, but also for its juice, decorative uses and *sharbat* qualities.

Origin and distribution

The fruit is believed to be native of Malaysia or Indonesia (Zewen and De wet, 1982), However, Pijpers *et al.* (1986) believed that it originated in India or Sri Lanka. It is now distributed around the world and cultivated mainly within 30° north and south of equator (Watson *et al.*, 1988). It is known in south eastern Asia for thousands of years (Watt, 1908). It was introduced to the USA by a Florida nursery over 95 years ago (Anon. 1887). In Queensland, Australia, it is grown for at least 100 years (George *et al.*, 1988) and is now one of the important and popular fruits of China, Taiwan, Thailand, USA, Brazil, Malaysia, Columbia, India, Pakistan, Indonesia and Myanmar.

Nutritive, medicinal and other uses

Carambola fruits are eaten fresh, used in salads, desserts and cooked as star-shaped slices. Carambola fruits are utilized for making preserve, candy, jam, chutney, pickle and ready to serve beverage (Singh, 2001). Fruits can also be dried, canned and frozen. The leaves have also been eaten as a substitute for sorrel. The fruit is a rich source of reducing sugars, ascorbic acid and minerals such as K, Ca, Mg and P. Ripe fruits of sweet carambola contain both oxalic acid (0.16%) and malic acid in quantities ranging from 1.0 per cent in unripe fruits to 0.51 per cent in ripe ones. Sugars present in both types consist largely of glucose with moderate quantities of fructose and traces of sucrose. The vitamin C content is also high (50 mg per 100 g) when compared with orange. Samson (1986) stated that the fruits contain several amino acids including serine, glutamic acid and alanine. The composition of carambola fruit is given in Table 3 (Watson *et al.*, 1988).

Table 3 : Composition of carambola fruits

Constituent	Content
Oxalic acid (g/100 g juice)	0.04-0.7
Acidity (meq/100 g juice)	1.9-13.1
pH	2.4-5.0
TSS (%)	5-13
Total sugar (%)	3.5-11.0
Water content (dry weight %)	90
Juice content (%)	60-75
Protein (%)	0.75
Crude fibre (%)	0.7
Vitamin A (IU/100 g)	560

Carambola is mostly consumed as dessert fruit in Australia. The fruits are also used for preparing juice, preserves, jam, jelly and pickles. In Philippines, the ground leaves are used as remedy for chicken pox, intestinal parasites and headaches (Coronel, 1983). Carambola fruit juice was found to contain a substance which retards oxidation of ascorbic acid. When orange juice containing 10% and 25% carambola juice was held for 2 weeks at 5°C, the retention of ascorbic acid was 83% and 91%, respectively, as compared to 77% in orange juice alone (Miller, 1955). The star sections of fruit are used in fruit salads and on cakes. The fruits have a wide appeal because of its decorative qualities.

The carambola fruit is made into preserves, jam, jelly and pickle. According to ayurveda, the ripe fruit of carambola produces a cooling effect and thus helps in curing diarrhoea, vomiting and dysentery. The ripe fruit contains 8.0% TSS, 0.567% titrable acidity, 10.85% total sugars, 0.7% protein and 13.8 mg per 100 g of vitamin C (Neog and Mohan, 1994). The acid fruits are often used for cleaning metal surfaces, especially for removing rust stains. Sweet type contains both oxalic and malic acid, whereas sour type contains only oxalic acid. Fruit contains 75 to 80% pulp, 5 to 7% total soluble solids, 0.5 to 1.0%acidity, 550 to 700 IU of the vitamin A and 0.3–0.5% minerals (Singh, 2001).

Pino *et al.* (2000) isolated volatile components of carambolas of optimum eating quality. Fifty-six components were identified (8.9 mg per kg), of which the major ones were butyl acetate, ethyl decanoate and hexadecanoic acid. Carambola fruit is a rich source of antioxidants (Shui and Leong, 2004 ; Suda *et al.*, 2005; Surinrut *et al.*, 2005).

Soil and climatic requirements

It can be grown on any type of soil ranging from sandy to heavy clay loam with adequate nutrition. However, it grows best on well drained, deep clay loams. It is tolerant to a range of pH levels, but 5.2 to 6.2 is preferred (Watson *et al.*, 1988). A good drainage is considered essential as it cannot stand waterlogging (Watson *et al.*, 1988).

The tree is well adapted under varying tropical and subtropical conditions and is suited to frost free areas. However, it prefers warm and moist condition. Areas with long hours of sunshine are preferred (more than 2000 h per year). Carambola grows best at temperature between 21 and 32°C and trees are seriously damaged if exposed for a long periods between -2 and -3°C and for short period at –4.4°C (Campbell and Malo, 1972). Annual rainfall between 1800 to 2500 mm is reported to be optimum for better growth and quality (Morton, 1987; Wahab Ngah *et al.*, 1989). In India, carambola is grown mostly on the west coast of India and on lower hill slopes of south India up to an altitude of about 1000 m. It is grown in almost all the frost free areas of India (Singh, 1969).

Cultivars

There are no standard cultivars of carambola in India, but two main types are distinguished, *i.e.* sweet and sour type (Singh, 1969, Singh, 2001). Diversity exists in carambola with respect to tree, leaf, leaflet characters, fruit size, shape, colour, taste, seed content and other biochemical traits. In India, organized plantation of carambola is rare, but few sweet or sour type carambola trees are easy to find in any garden throughout the country. Srivastava (2005) evaluated four carambola genotypes, namely Malviya Carambola-1, Malviya Carambola-2, Malviya Carambola-3 and Malviya Carambola-4 under Varanasi conditions of India and found Malviya Carambola-1 as the best genotype due to high fruit set, minimum fruit drop, longer shelf-life, high total soluble solids, ascorbic acid and total sugars. The physico-chemical composition of fruits of Malviya Carambola-1 revealed creamy yellow colour, average fruit weight (66.79 g per fruit), 8.1%

TSS, pH 3.98, high ascorbic acid content (16.98 mg per 100 g pulp), low seed (1.03 per fruit) and fibre content (0.31%).

The sweet types have low acid (0.4%) with 5% sugars and the sour types contain as much as 1% acid. Some Chinese types like Fuang Tung are very sweet and Brazilian ones are rich in vitamin C. Some superior types of carambola are available at Columbia (Icambola), Taiwan (Tean Ma, Min Tao) and Hawaii (Golden Star). The cultivars of carambola commonly grown in different countries are listed in Table 4.

Table 4 : Carambola cultivars grown in different countries

China	:	Ell-Lin, Fuang Tung, Xiangmi Yangtao, Cheng Chwei, Ching Chyan
Florida, USA	:	Golden Star, Fwang Tung, Dah Pon, Tean Ma, Arkin Maha, Star King, Thai Knight, Wheeler, Newcomb
Brazil	:	IPA25.1, IPA20.3, IPA7.2, IPA1.3, IPA22.3, Tean Ma, Fwang Tung, Arkin, Malaysia, Golden Star, Nota-10
Singapore	:	Leng Bak, Jurong
Taiwan	:	Iu Tho, Cheng-Tsey, Dah Pon, Er-lin, Soft Sih
Malaysia	:	B1, B2, B4, B6, B8, B10, B11, Daguotianyangtao
Australia	:	Giant Sam, Jungle Gold, BCP-1, Hosie and Chujuba
Thailand	:	Fwang Tung, Thai Kmgut

Wang *et al.* (1997) found the fruits of cultivars Cheng Chwei and Ching Chyan most suitable for export on account of their good quality determined by sensory evaluation of surface, colour, flavour and texture and saleability after simulated export storage, while fruits of cultivars Tainoung No.1 and Malaysia were considered unsuitable for export due to their high degree of mechanical (30%) and chilling injury (100%) after simulated export storage. Teixeira *et al.* (2001) analysed physical and chemical characteristics of 6 carambola cultivars, *viz* Tean Ma, Fwang Tung, Arkin, Malaysia, Golden Star and Nota-10 and reported that Arkin, Tean Ma and Nota-10 were the largest, heaviest and had the greatest total soluble solids and soluble sugar contents and a good ratio of total soluble solids to total titratable acidity. In China, the cultivar Ell-Lin was found most suitable for making juice, preserves and for canning, and cultivar Waiwei is used for preparation of jam (Yu and Wang, 1987). Prati *et al.* (2002) found sweet carambolas better than the sour ones for processing canned fruit due to superior texture and flavour.

Propagation

Carambola is often propagated through seed. The seeds have low viability and hence should be sown fresh. After cleaning and drying the seeds, they should be sown in pots and regularly watered. The seedlings are transplanted to individual pots or plastic bags and allowed to harden. They are then translplanted to the field. One year old seedlings are used as rootstocks.

Asexual propagation of carambola is easily done by grafting. Soft wood grafting using cleft method on one year old seedling rootstocks raised in polythene bags measuring 25 X 15 cm filled with soil:FYM mixture in 2:1 ratio gave cent per cent success during February under Lucknow conditions in India. The other techniques of grafting in common use are wedge, splice (whip and tongue), side veneer and approach grafts (Watson *et al.,* 1988). Approach grafting is, however very successful (Chua and Teoh, 1973). Campbell *et al.* (1985) suggested forcing of bud wood by removing leaves of the scions before 3 to 7 days of budding. Veneer grafting is normally practised for propagation of carambola in Florida. Inarching and layering also gave cent per cent success (Naik, 1949).

Cent per cent success was also achieved by air layering of one year old terminal shoots which were ringed (1.5 cm wide) and treated with IBA at 10000 ppm at the time of layering (Banarjee *et al.,* 1980). Borah and Das (2000) achieved success in propagating carambola by stem cuttings treated with IBA. Bastos *et al.* (2004) found top cuttings of carambola, treated with IBA solutions at 0, 1000, 3000, 5000 or 7000 ppm for 10 seconds, resulted in greater survival (49.26%), rooting (34.84%), and number of medium-sized roots per cutting (20.51) than the base cuttings.

Planting and after-care

One year old healthy grafts or seedlings are planted in prepared pits of 1 x 1 x 1 m. The pits are filled with top soil and farm yard manure in 3:1 ratio and left to decompose for 2 to 3 months before planting. The plantation can be done either in July-August or February-March. The planting distance for carambola should be kept 8 X 8 m or 6 X 6 m depending upon the growth habit, vigour of the cultivar and fertility of soil. The planting distance can further be reduced to 6 X 3 m or 3 X 3 m for accommodating higher number of trees per unit area.

Carambola is a gross feeder and fruit size is increased with good nutrition and watering. A dose of 10 kg well rotten farm yard manure (FYM), 100 g nitrogen, 50 g phosphorus and 100 g potassium should be applied to a one year old plant. This dose should be increased till 10 years of tree age. Thereafter, a constant dose of 100 kg FYM, 1 kg nitrogen, 0.5 kg phosphorus and 1 kg potassium should be applied every year. The dose should be applied in two equal split doses, first in July-August and second in January-February to ensure quality production of fruits. Application of other macro and micronutrients such as calcium, magnesium, boron, zinc, copper improves fruit quality. Foliar sprays of $ZnSO_4$ (1%) proved effective in quality improvement.

The tree has high water demand. It is advisable to irrigate the plants at 10 to 15 day intervals during summer and at 20 to 25 day intervals during winter. Mulching is used particularly in young plantations to check the weed growth and conserving soil moisture around the tree. It is beneficial for growth as well as for reducing water demand. In Australia, organic mulching with peanut shells or straw is in use for control of weeds. The application of a mixture of 80% diuron (1.5 kg a.i.) and paraquat (1.5 litre a.i.) in 400 litre of water per ha is also effective in controlling weeds.

Pruning of low hanging, unthrifty and criss-cross branches is desirable. Young trees are trained retaining four to six sturdy wide crotch angled branches to build the tree framework. The more vigorous, upright narrow crotch angled types may be trained to a vase or palmette system. The vigorous top leading branches should be shortened back to suitable side branches at regular interval. Campbell (1989) suggested pruning of main stem when it grows vigorously without producing side branches, which will encourage new side branches to grow. He also suggested regular pruning of terminal branches to a height of 3.5 to 4.5 m when the trees are 5 to 6 m tall.

Carambola trees grow and bear fruits round the year in their native tropical environment. Fruits are borne mainly on the canopy periphery, and the delicate skin is easily damaged during windy conditions (wind-scar). Pruning, as well as fruit drop caused by strong winds, stimulates re-bloom of carambola trees. In subtropical southern Florida, cool temperatures and dry winds during the winter inhibit tree growth and shorten the production season to July-February. Manual, selective pruning and crop removal improves yield and quality of fruits and regulate flowering and fruiting.

Harvesting and yield

Seedlings take about 4 to 5 years to come in to bearing but grafted trees usually start bearing after 2 to 3 years of planting. After maturity the fruits remain on the tree hardly for a week and then drop off (Nand, 1970). The fruits should be harvested as soon as they mature. Peng Fatt and Hassan (1984) suggested that harvesting should be done when 25 to 75% of surface has turned yellow. They suggested the following colour chart for harvesting of carambola, of which stage 3 was recommended for harvesting:

Stage	Colour
1	Green
2	Yellow tinge over 25% of surface
3	25-75% of surface yellow
4	75-100% of surface yellow
5	Fully yellow or orange

Fruit should be picked carefully and preferably by hand rather than a pole and bag, in order to prevent damage and bruising to the fruit.

The yield varies from 20 to 50 kg on a 2-year old tree to 400 to 500 kg on a six year old tree. The average yield has been mentioned as 50 to 100 kg per tree in India (Singh, 1969; Mehrotra, *et al.*, 1996). In South Africa, the trees usually come into bearing after 3 years and yield of 50 kg per tree has been obtained after 5 years; after 8 years yield may reach 100 to 150 kg per tree.

References

Anonymous (1887) Catalogue Royal Palm Nursery, USA.

Banerjee, D.P., Chatterjee, B.K. and Rao, D.P.(1980) *Indian Agric.*, **24** : 165-168.

Bastos, D.C., Martins, A.B.G., Scaloppi,J.E.J., Sarzi,I. and Fatinansi, J.C. (2004) *Revista Brasileira de Fruticultura,* **26**(2): 284-286.

Borah, A., Das, R.P. (2000) *Haryana Journal of Horticultural Sciences,* **29** (3-4) : 196-198.

Campbell, C.W. (1989) *Proc. Int. Amer. Soc. Hort. Sci.,* **33 :** 66-71.

Campbell, C.W. and Malo, S.E. (1972) Fruit Crops Fact Sheet, Florida Coop. Extension Service (FC-12) p. : 2.

Campbell, C.W., Knight, R.J. and Olszack, R. (1985) *Proc. Int. Amer. Soc. Hort. Sci.,* **98** : 145-149.

Chua, S.E. and Teoh, I.S. (1973) *Singapore J. Primary Industry,* **1** : 87-95.

Coronel, R.E. (1983) Promising Fruits of the Philippines, College of Agriculure, University of Philippines, Los Banes p. : 508.

George A.P., Nissen, R.J. and Watson, B.J. (1988) *Maroochy Hort. Res. Stn. Rept.*, No. 5 pp. 119-120.

Mehrotra, N.K., Singh, H. and Singh, T. (1996) *Indian Hort.*, **41** (3) : 10-11.

Miller, E.V. (1955) *Abstr. Pl. Physiol.*, **30** (suppl) : XXXVII.

Morton, J.F. (1987) In : Fruits of Warm Climates, (Ed. Julia F. Morton), Creative Resource Systems, Winterville, FL., pp. 125-129.

Naik, K.C. (1949) South Indian Fruits and their Culture, P. Varadachary and Co., Chennai.

Nand, D. (1970) *Indian J. Hort.*, **27** : 147-152.

Neog, M. and Mohan, N. K. (1994) *Indian Hort.*, **39**(2) : 28-31.

Peng Fatt, L. and Hassan, A. (1984)*Teknology Makanam*, Jil, 3, Bil. **2** : 39-44.

Pijpers, D., Constant, J.G. and Jansen, K. (1986) The complete book of fruit, pp. 70-71.

Pino, J.A., Marbot, R. and Aguero, J. (2000) *Journal of Essential Oil Research*, **12** (4) : 429-430.

Prati, P., Nogueira,J.N. and Dias,C.T.S. (2002) *Boletim do Centro de Pesquisa-e-Processamento de Alimentos*, **20**(2): 221-246.

Samson, J.A. (1986) Tropical Fruits, Longman Sci and Tech.

Shui, G.H., Leong, L.P. (2004) *Journal of Chromatography*, **1022**(1/2): 67-75.

Singh, I.S. (2001) *Indian J. Hort.*, **58** (1&2) : 178-182.

Singh, R. (1969) Fruits, National Book Trust, India, pp. : 126.

Srivastava, K.K. (2005) *Orissa Journal of Horticulture*, **33**(2): 20-25.

Suda, I., Oki,T., Nishiba,Y., Masuda, M., Kobayashi, M., Nagai, S., Hiyane, R. and Miyashige, T. (2005) *Journal of the Japanese Society for Food Science and Technology*, **52**(10): 462-471.

Surinrut, P., Kaewsutthi,S.and Surakarnkul,R. (2005) *Acta-Horticulture*, **679**: 201-203.

Teixeira,G.H. de A., Durigan, J.F., Donadio, L.C. and Silva, J.A.A. da (2001) *Revista Brasileira de Fruticultura*, **23**(3) : 546-550.

Wahab Ngah, A., Ahmad, I. and Hassan, A. (1989) *Proc. Int. Amer. Soc. Trop. Hort.*, **33** : 30-43.

Wang, T.C., Shiesh, C.C. and Lin, D.S. (1997) *Journal of Agricultural Research of China*, **46** (3) : 278-293.

Watson, B.J., George, A.P., Nissen, R.J. and Brown, B.I. (1988) *Queensland Agric. J.*, **114** : 45-51.

Watt, G. (1908) In : The Commercial Products of India, John Murray, London

Yu, R.C. and Wang, W.C. (1987) *J. Agric. Res.*, China, **36** : 196-206.

Zewen, A.C. and De Wet, J.M.J. (1982) Dictionary of cultivated plants and their centres of diversity, Centre for Agricultural Publishing and Documentation, Wageningen.

Chapter 4 Custard apple

Botanical Name	:	*Annona squamosa* L.
Regional/ Vernacular Names	:	Sugar apple, sweetsop, ata, seetaphal, attichakka, seethaapazham, ato, seethaapholo, gandhagataram, seethaphalam, gandhagaalaramu, seetaaphalamu
Family	:	Annonaceae

Custard apple fruit is formed by fusion of pistil and receptacle into a large fleshy aggregate fruit. The tree, due to hardy nature, has wider adaptability to soil and climatic conditions and freedom from pests. As a result, the custard apple has become naturalized in tropical and subtropical parts of the country, with a tremendous scope for further expansion.

Origin and distribution

Custard apple is a native of the West Indies and south America (Bailey, 1949). In India, only the sweetsop or sugar apple (*A. squamosa*) is commercially grown. Custard apple plants are found growing wild in forests in Andhra Pradesh, Tamil Nadu, Assam, Chhattisgarh, Madhya Pradesh and Uttar Pradesh. It is a commonly cultivated fruit in India, China, Philippines and Cuba. The fruit has commercial importance in Egypt and Central Africa.

In India, the fruits are very poplar in the Deccan Plateau and also grown commercially on a smaller scale in Assam, Andhra Pradesh, Bihar, Karnataka, Madhya Pradesh, Maharashtra, Orissa, Rajasthan, Gujarat and Uttar Pradesh. Andhra Pradesh appears to be a leading state in custard apple production in India. In recent years, area under custard apple has increased manifold in Andhra Pradesh and Maharashtra.

Nutritive, medicinal and other uses

Custard apple is mostly consumed as table fruit. It is also used in ice creams and preserved as jam, jelly or other products on limited scale.

Directly baked fruits are commonly eaten in Andhra Pradesh (Rao, 1974). The edible portion of fruit or pulp is creamy or custard like, granular with a good blend of sweetness and acidity. The pleasant flavour and mild aroma have universal liking. The leaves are known to have insecticidal properties. The nutritive value of the fruit has been given in Table 5.

Table 5: Composition of custard apple fruit (per 100 g of edible portion)

Constituent	Content
Moisture (%)	73.5
Carbohydrates (%)	23.9
Protein (%)	1.6
Fat (%)	0.3
Calcium (%)	0.02
Phosphorus (%)	0.04
Iron (mg)	1.0
Vitamin C (mg)	37

The fruits are rich in carbohydrates. They are composed of peel, pulp and seed in varying proportions. The immature fruits, seeds, leaves and roots are of considerable medicinal value both in ayurvedic and unani systems of medicine. The seeds are abortifacient and roots are drastic purgative. The seeds contain about 30% oil which can be used in soap and paint industry. The seed cake containing about 4% nitrogen is beneficial as manure. Its insecticidal properties have also been demonstrated against *Nephotettix virescens* (Mariappan and Saxena, 1984).

Soil and climatic requirements

Custard apple is a woody shrub or tree with shallow root system not requiring deep soils. Trees occur naturally in rocky, gravelly, well-drained sandy loam soil. However, they may grow well in arable, red, sandy shallow soils slightly acidic in reaction. Well-drained open textured soils with pH range from 5.5 to 6.5 are preferable. Heavy soils are not preferred, especially in waterlogged areas. It is tolerant to many soil conditions like presence of high chlorine, rockiness, shallowness, light and heavy soils with good drainage, but saline and alkali soils are not suitable for custard apple. Plants are capable of growing in as much as 50% lime in the soil and 300 ppm of chlorine in irrigation water. Since, waterlogging causes tree decline, heavy soils with poor drainage, sub-soil hard pan or high water table are to be avoided.

It is a tropical plant but can be grown in mild sub-tropics also. It grows well from sea level to about 1000 m altitude and from few centimeters of rainfall in Rajasthan to 250 cm in Sri Lanka. However, average rainfall of 50 to 75 cm is optimum. Although custard apple withstands heat and drought conditions, high atmospheric humidity is necessary during flowering to improve fruit set. Continuous rains during fruit set are not desirable. It cannot withstand frost or long cold period. Fruits become hard and do not ripen in severe cold. They are evergreen but when grown in north India, they often shed their leaves in the cold weather. Hot dry wind at the time of flowering in late spring results in poor fruit set in north India. In tropics, water stress puts this plant in quiescence, while in subtropics low temperature and water stress in summer creates these conditions which help the plants to tolerate drought in a better way.

Cultivars

Custard apple is often propagated by seed. There are a few standard cultivars grown in different parts of the country. Most of the custard apple types are considered to be bud sports or variants originated from the vast seedling populations. These types have been named after their colour, place of origin, nature of areole, etc. In India, collection, selection and evaluation of custard apple varieties have been carried out in Andhra Pradesh and Maharashtra. The important cultivars of custard apple are Balanagar, Washington 98797, Washington 107005, Red Sitaphal, Local Sitaphal, Mammoth, Barbados, British Guinea, Brandy and Islander. The cultivar Balanagar has better fruit quality as compared to others. Arka Sahan, a hybrid developed by Indian Institute of Horticultural Research (IIHR), Bangalore has also become very popular. The chief characteristics of important cultivars are given below:

Balanagar

This is a promising selection made from the Mahboobnagar District of Andhra Pradesh. Fruit shape is pyramidal or chordate, areoles very rough, pitted, usually remains green when ripe. Flakes crisp and slightly leathery; pulp bright white, very sweet, juicy with medium granular mesocarp; flavour excellent. The tree bears 50 to 60 fruits per tree and the average seed content is 25 per fruit.

Washington 107005

This variety was introduced from Australia. Tree is vigorous with broad and large leaves. Fruits medium to large size, spherical to short conical in shape. Rind greyish green with fine bloom. Flakes soft, moderately adherent to seed. Pulp bright white, sweet and buttery. The tree bears 70 to 80 fruits on an average.

Arka Sahan

Arka Sahan is a hybrid resulting from a cross between Island Gem (*Annona atemoya*) X Mammoth (*A. squamosa*). Arka Sahan fruits are ready for harvest in September-October and mature fruits (210 g) take about 6 to 7 days to ripe. The skin has a waxy bloom, light green in colour, moderately thick (0.5 cm) with large, flat-eyes. The creamy white flesh is juicy with mild pleasant aroma and tender with few seeds (9 per 100 g) and large segments. The edible pulp is remarkable for its sweetness with 22.8 per cent total sugars and TSS of 30^0Brix or more as against 24°Brix in Mammoth. One hundred gram of pulp contains 2.49 g of crude protein, 42.29 mg P and 225 mg Ca. An average fruit yield of 12 tonnes per ha can be obtained (Maurya and Singh, 2004).

Propagation

Custard apple is propagated by both sexual and asexual methods. The seeds viability lasts for 3 to 4 years. Fresh seeds germinate in 20 to 30 days giving as high as 90 per cent germination. Treating seeds with 500 ppm GA_3 improves germination. Seeds are sown in beds or polybags filled with soil, the latter is preferred because of ease and low mortality during transplanting. The 30 cm tall seedlings become ready for transplanting.

Custard apple can easily be propagated by shield, patch, modified forkert and chip budding in early spring on the commencement of sap flow or in the autumn. If budding is done during active growth season, the excessive sap flow may hinder bud union. Dormant last year's shoots serve well as bud wood. Grafting was found superior to budding for propagating custard apple in terms of per cent success and subsequent growth.

The whip, splice, tongue, cleft, crown, veneer and inarching methods of propagation have been successful in custard apple. The grafting should be done in early spring on one year old mature scion and one year old rootstock. Cleft grafting during spring season on one year old rootstocks raised in polythene bags gave 64 to 92% success under

Lucknow conditions (Shukla and Ali, 2007). Soft wood grafting was found most successful in Akola, Maharashtra in terms of success (87.02%), survival (73.73%), shoot length (37.08 cm) and number of functional leaves (12.10) when grafted during February on 12-month old rootstocks (Pawar *et al.*, 2003). Soft wood grafting in March was found to be the best method with 51% graft take for custard apple under Anantpur conditions in Andhra Pradesh (Pareek and Vishal Nath, 1996). Mortality during transplanting in custard apple can be reduced by planting in dormant season, avoiding root damage, excessive watering and improper fertilization.

Normally, seedlings of the custard apple are used as rootstocks, but *Annona reticulata* (Ramphal) has proved to be a better rootstock for *A. squamosa*, imparting longer life, better tree vigour, fruit yield and quality to the scion variety. Pawar *et al.* (2003) obtained significantly higher shoot length, shoot diameter and the number of functional leaves in custard apple seedling rootstocks raised from seeds sown in February than those from seeds sown in March and April in Akola, Maharashtra.

Planting and after-care

Pits of 60 x 60 x 60 cm size are dug and left open to sun for a week. They are filled with top soil mixed with 25 to 30 kg FYM, 100 g each of urea, single super phosphate and muriate of potash at least a fortnight prior to planting. Planting is done in the rainy season at a distance of 5 x 5 m in square system. In order to secure better pollination, closer spacing promotes better fruit set. Now-a-days, high density plantation of custard apple is gaining popularity in Maharashtra and other parts of the country.

Custard apple is mostly cultivated on poor soils, the need for manure and fertilizer application becomes essential. Besides this, proper manure and fertilizer application checks decline and extend longevity of trees. The area below the tree canopy should be weed free. Fertilizer application should be done during rainy season in the basin but not within 30 cm around the trunk. At Rahuri, application of 250 g nitrogen, 125 g phosphorus and 125 g potassium per plant per year gave higher yield (Anon, 1983). Pareek and Vishal Nath (1996) reported that application of nitrogen @ 250 g per tree per year increased productivity of custard apple, while phosphorus and potassium failed to give any response. Rao (1974) reported that malnutrition and lack of moisture increase number of stone fruits which do not ripen properly. Mondal and Chattopadhyay (1993) reported that increase in application of NPK from 120 g each per plant to 240 g each per plant annually for two consecutive years produced higher number of flowers, better fruit set, retention and yield.

Custard apple is considered drought tolerant. It produces good crop even without any irrigation. Fruit set and development occur during rainy season, which coincides with the period of ample moisture in the soil and humidity in the atmosphere. Two to three irrigations before the onset of monsoon to promote fruit set and one or two irrigations after the monsoon to increase fruit size are recommended. Custard apples are mostly grown as rainfed crop. In rainy season, efforts should be made to conserve water in rootzone.

In initial years of planting, economic returns can be obtained from the fruit tree interspaces by cultivating intercrops. Groundnut, smaller millets, linseed, pea, gram, radish, tomato, brinjal, chilli and other small vegetables can be grown successfully in custard apple orchards. Custard apples are often grown on uneven sandy or rocky soils, which are not fit for clean cultivation. But, the farmers around Lucknow (Uttar Pradesh) often cultivate brinjal and other vegetables in the most commonly prevalent seedling orchards of custard apple.

Leaf shedding during winter season and production of floral buds on new shoots indicate the necessity of light pruning. In order to build up a strong framework, only a few scaffold limbs are allowed to develop in a well distributed manner, and other branches are removed. Subsequently, very old and weak branches are thinned out from time to time giving proper space to new and young branches. Pruning should, however be carried out when the plant is about to put forth new growth in early spring after dormancy. The completion of leaf fall is essential for the initiation of new growth in custard apple. Use of two sprays of Potassium Iodide, a chemical defoliant @ 0.5 per cent at weekly interval gives early and prolonged harvest with higher yield.

Since large number of custard apple plants are of seedling origin and have uncertain yield, rejuvenation of seedling plants through top working using commercial cultivars can play a vital role in achieving higher and quality production. The seedling trees can be top worked through cleft or bark grafting using two or three scions on large limbs at the end of the dormant season. After grafting, proper development of tree framework is necessary to achieve quicker results from top working.

Harvesting and yield

Custard apple has a short gestation period and starts flowering after 3 years of planting. It bears flowers on the current season's growth and rarely on older branches. The flowering period is very long commencing from March-April, continuing up to July-August and peak flowering period takes place in April and May. An average plant bears about 2000 flowers and the fruit set is about 2 to 3 per cent which can be increased by hand pollination with suitable parent. Fruit set is better when humidity is high. A plantation with more than one cultivar is desirable for ensuring better fruit set and yield. For hand pollination, flower should be collected at about 4 pm. On pollination, pollen tube reaches the embryo sac without any barrier and fertilization takes place normally. The cultural practices including close planting, regular irrigation during summer, overhead spraying of water and mulching help in enhancing the setting of fruits. Custard apple fruits take about 4 months to mature from the time of anthesis. The period is reduced, when the soil moisture is optimum.

The fruits are examined individually every alternate day to determine the stage of maturity for harvesting. The fruits are handpicked. The peak period of ripening is beginning of October to November. The following three maturity indices are usually adopted in custard apple.

(i) The light green fruit colour.

(ii) Yellowish colour between the carpels.

(iii) Initiation of cracking of skin between the carpels.

The size of fruit is determined by number of seeds, which depend on the level of pollination. The more the number of seeds in a fruit, the higher is the fruit weight and size. Custard apple starts bearing at the age of 3 years and in the fourth year, it gives as many as 50 fruits, while older plants may yield about 100 to 150 or more fruits, depending upon the size of the tree.

References

Anonymous (1983) Report, All India Coordinated Fruit Improvement Project, pp. 18-19.

Baily, L.H. (1949) Manual of Cultivated Plants, Macmillan Co., New York, p. 419.

Mariappan, V. and Saxena, R.E. (1984) *J. Econ. Entom.*, **77**: 519-21.

Maurya, I.B. and Singh, D.K. (2004) Advances in Arid Horticulture, Vol.II (Eds. P.L. Saroj and O.P.Awasthi), IBDC,Lucknow, pp. 129-141.

Mondal, A. and Chattopadhyay, P.K. (1993) *J. Pot. Res.*, **9** : 375-379.

Pareek, O.P. and Vishal Nath (1996) Coordinated Fruit Research in Indian Arid Zone : A Two Decade Profile (1996-1995), pp.67-74.

Pawar, D.M., Ingle, V.G. and Panchabhai, D.M. (2003) *Annals of Plant Physiology*, **17**(1): 53-55.

Rao, S.N. (1974) *Indian Hort.*, **19** : 19-21.

Shukla, S.K. and Ali, A. (2007) *CISH News*, **8**(4) : 2-3.

Venkatratnam, L. (1973) Horticulture in Central India, Farm Information Unit, Directorate of Extension, Ministry of Agriculture, New Delhi.

Chapter 5 Wood apple

Botanical Name	:	*Feronia limonia* (L) Swingle syn. *Feronia elephantum* Correa or *Limonia acidissima* L.
Regional/ Vernacular Names	:	Bilin, kait, kaitha, kavitha, katbel, kavit, kotha, kothi, kothun, kavatha, kavith, kovit, velaga, vilanga, bela, vila, vilatti, elephant apple
Family	:	Rutaceae

Wood apple is a thorny tree commonly found in dry deciduous forests. The tree is very hardy and has tolerance to drought and salinity. Its cultivation as a fruit tree is rare, but fruits of naturally occurring trees in community lands, forests and on roadsides are commonly used for chutney and pickle making. Fruits being rich in pectin, minerals and vitamins offer immense scope to utilize them for processing.

Origin and distribution

The wood apple is native and common in the wild in dry plains of India and Sri Lanka and grown along roads, on field bunds and occasionally in orchards. In India, the fruit was traditionally a "poor man's food" until processing techniques were developed in the mid-1950's. It is one of the very hardy trees found growing all over the plains of northern, central, eastern and southern regions of India, especially in the semi-arid and arid regions. It is more common in Deccan, Maharashtra, Madhya Pradesh, Uttar Pradesh, Chhatisgarh, Bihar and Jharkhand, it also occurs in the western Himalaya up to an elevation of about 500 m. It is rare to find a well planted wood apple orchard. The information regarding area and production under this fruit crop in India is not available. Sporadic trees are found in forests, community land, on field bunds and road sides.

Nutritive, medicinal and other uses

The nutritional and economic importance of wood apple has not been widely recognized. This fruit can be exploited for regular cultivation in dry areas. Commercial cultivation requires support from processing industry. It makes excellent chutney and can be used as adjunct in jelly and also made into squash. Being rich in pectin, it makes good jelly. Pulp contains 2.66% pectin on fruit weight basis (Roy and Mazumdar, 1988). The pulp is sweetened with *gur* or sugar and eaten fresh. The ripe fruits contain sour-sweet, aromatic pulp which is about 70 per cent of total weight. Wood apple pulp contains significant amount of protein. The fruit is a rich source of riboflavin, acidity (2.3 per cent) and sugars (7.2 per cent). Wood apple fruits available during winter contain 40 to 50% pulp with 14 to 18% total soluble solids (TSS), 3 to 4% acidity, 1.6 to 2% pectin and they are very rich in calcium, phosphorus and riboflavin (Singh, 2001) and they may be utilized for making excellent quality of the jelly, chutney, ready to serve beverage and squash. Wood apple fruits had higher content of both acidity and pectin, hence most suited for jelly making (Singh *et al.*, 1999). The composition of wood apple fruit is given in Table 6 (Anon, 1956).

Table 6 : Composition of wood apple fruit (per 100 g of edible portion)

Constituent	Content	Constituent	Content
Moisture (g)	64.2	Copper (mg)	0.21
Protein (g)	7.1	Manganese (mg)	0.18
Carbohydrates (g)	17.0	Zinc (mg)	0.46
Fat (g)	0.3	β-Carotene (μg)	61
Total minerals (g)	0.3	Thiamine (mg)	0.04
Calcium (mg)	4	Riboflavin (mg)	0.17
Phosphorus (mg)	9	Niacin (mg)	0.8
Iron (mg)	0.5	Vitamin C (mg)	3
Magnesium (mg)	41		

The pulp (100 g) from ripe fruit contains moisture (74.0%), protein (8.00%), fat (1.45%), carbohydrates (7.45%), ash (5.0%), calcium (0.17%), phosphorus (0.08%), iron (0.07%), tannins (1.03%), while seeds contain 4.0% moisture, 26.18% protein, 27% fat, 35.49% carbohydrates, 5.03% ash, 1.58% calcium, 1.43% phosphorus, 0.03% iron, 0.08% tannins. The seeds contain oil, high in unsaturated fatty acids. The rind must be cracked with a hammer. The scooped-out pulp though sticky, is eaten raw with or without sugar, or is blended with coconut milk and palm-sugar syrup and drunk as a beverage or frozen as an ice cream. It is also used in chutneys and for making jelly and jam. The jelly is purple and much like that made from black currants. Bottled nectar is made by diluting the pulp with water, passing through a pulper to remove seeds and fibre, further diluting, straining, and pasteurizing. A clear juice for blending with other fruit juices, is obtained by clarifying the nectar with Pectinol R-10. Pulp sweetened with syrup of cane or palm sugar, may be canned and sterilized. The pulp can also be freeze-dried for future use. The pectin has potential for multiple uses in pectin-short India, but it is reddish and requires purification.

Medicinal properties of wood apple have been dealt with in *Charak Samhita* and *Sushruta Samhita,* two early medical treatises in Sanskrit. It has great demand from native system of ayurvedic medicines. Fruit is used as stomachic and stimulant while leaves are carminative. The fruit is used in India as a liver and cardiac tonic, and when unripe, as an astringent means of halting diarrhoea and dysentery and effective treatment for hiccough, sore throat and diseases of the gums. Pulp is poulticed onto bites and stings of venomous insects, as is the powdered rind (Kirtikar and Basu, 1935). Maiti and Mishra (2000) reported antivenom activity in wood apple. The Bark is prescribed for biliousness. Juice of young leaves is mixed with milk and sugar candy and given as a remedy for biliousness and intestinal troubles of children. The powdered gum, mixed with honey, is given to overcome dysentery and diarrhoea in children. Oil derived from the crushed leaves is applied on itch and the leaf decoction is given to children as an aid to digestion. Leaves, bark, roots and fruit pulp are all used against snakebite. The spines are crushed with those of other trees and an infusion taken as a remedy for menorrhagia. The bark is chewed with that of *Barringtonia* and applied on venomous wounds. Sheeja *et al.* (2005) justified the use of wood apple by the tribal people as a substitute for bael (*Aegle marmelos*) and *vice-versa* due to the similar composition of the leaf extracts to some extent.

The wood is yellow-grey or whitish, hard, heavy, durable, and valued for construction, pattern-making, agricultural implements, naves of wheels, oil crushers, house building, rollers for mills, carving, rulers, and other products. It also serves as fuel. The trunk and branches exude a white, transparent gum especially after the rainy season. It is utilized as a substitute for, or adulterant of, gum arabic, and is also used in making artists' watercolors, ink, dyes and varnish. It consists of 35.5% arabinose and xylose, 42.7% *d*-galactose, and traces of rhamnose and glucuronic acid. The hard dry shells of the fruits are made into snuff boxes (Bhasin *et al.*, 1958). The leaves are eaten by sheep and goat. The leaves have an aroma of aniseed and yield 0.73 per cent essential oil.

Soil and climatic requirements

Wood apple can grow in a variety of soils, including marginal and degraded soils of arid and semi-arid regions. It also tolerates salinity to a certain extent (Hebbara *et al.*, 2002). The tree shows a marked preference for black cotton soils for its optimum growth and fruiting. As regards climatic requirements, it is mostly grown in dry regions of subtropical and tropical regions of the country. It does not grow well above 1500 m. The mature plants can tolerate low temperature (0 to 15°C) as well as a temperature as high as 47.5°C (Troup, 1921). The warm season appears conducive for the initiation of floral buds. The tree sheds its leaves and the branches are bare for a short period in the cold season during January and the flowering starts in February-March. It possesses great tolerance to drought.

Cultivars

A lot of diversity is available in natural population of wood apple with respect to tree and fruit characters. There are two types, one with large and sweetish fruits, while the other with small and acidic fruits. Singh *et al.* (1999) studied the physico-chemical variability in wood apple fruits, grouping them into two categories, *i.e.* oblong and round. Oblong group had small sized fruits with low content of seed and fibre (Table 7).

Table 7 : Variability in physico-chemical traits of wood apple fruits

Characters	Oblong type	Round type
Fruit weight (g)	117-162	155-250
Pulp (%)	40.5-48.6	46.2-49.8
Seed (%)	3.8-5.5	7.6-7.8
Shell (%)	44.6-50.6	37.8-41.7
Fibre (%)	2.9-3.2	4.5-4.8
Total Soluble Solids (^{0}Brix)	14.3-15.3	15.6-16.2
Acidity (%)	3.9-4.2	3.1-4.0
Pectin (%)	1.6-1.7	1.8-2.0
Vitamin C (mg/100 g)	5.2-6.0	5.3-6.6

A survey was conducted in Lucknow and Jhansi districts of Uttar Pradesh; Tikamgarh and Datia districts of Madhya Pradesh to identify the promising wood apple genotypes with regular and prolific bearing, disease and pest free healthy trees and thin-shelled fruits of good size and shape (Shukla *et al.*, 2006). Seventeen genotypes were collected and screened on the basis of physico-chemical traits. Fruit weight ranged in between 231.6 g and 641.66 g, while TSS varied from 9.4 to 16.3^{0}Brix. Fruit length from 6.4 to 9.5 cm, fruit breadth from 5.9 to 10.1 cm, fruit volume from 227.93 to 671.37 cc, acidity from 0.83 to 2.76%, vitamin C from 1.68 to 3.04 mg per 100 g pulp, total sugars from 3.46 to 5.64%, shell thickness from 0.33 to 0.6 mm and shell weight from 100.00 to 161.66 g were recorded in different accessions. Analysis of mature fruits gave 120 (48.97%) to 366.66 g (71.6%) pulp per fruit.

Shete *et al.* (1999) evaluated fifteen local strains of wood apple from Rahuri, Maharashtra for physico-chemical characters. The weight of the fruit, pulp content and rind thickness varied from 173 to 540 g, 58.0 to 75.93 per cent and 0.2 to 0.5 cm, respectively and the total soluble solids, acidity and total sugars ranged from 14 to 18.5 per cent, 1.04 to 4.5 per cent and 4.08 to 8.47 per cent, respectively among the genotypes described. It appears that the organoleptic quality of wood apple pulp depends upon the brix:acid ratio. Higher values make the fruit pulp more palatable. The evaluation suggested that the superior strains/ genotypes No.5, No.9, No.10 and No.15 hold promise for future cultivation. The Marathwada Agricultural University (MAU), Parbhani, Maharashtra released a type called HB-10 having large-sized fruits. Average weight of fruit is 350 g with weight of pulp 224 g (Karale, 2002). Shukla *et al.* (2007) screened wood apple genotypes and found CISH WA-4, CISH WA-10 and CISH WA-16 promising.

Propagation

Wood apple is often propagated by seed. Freshly extracted seeds should be used for raising seedlings. Seeds sown fresh usually show a high percentage of germination (66%). About 31,000 plants can be raised from a kilogram of seed (Troup, 1921). Seedlings are transplanted a year later. Polyembryony to the extent of 63 per cent was observed in wood apple (Karale, 2002). Polyembryonic seedlings are vigorous in growth and breed true to type. Sappandi *et al.* (2006) studied seed germination and viability of wood apple seeds. Seed soaking in water for 24 hours or treatment with GA_3 @ 400 ppm for five minutes enhanced seed germination. Seeds lose their viability after extraction during storage, when the storage period enhanced from 15 to 150 days.

Wood apple can be propagated successfully by soft wood/ cleft grafting on seedling rootstocks. It is being successfully propagated by soft wood grafting in Uttar Pradesh, Karnataka and Gujarat. Hiwale (2006) standardized the method of soft wood grafting in wood apple at Godhra, Gujarat. The scion shoots defoliated 10 days before grafting were grafted on one year old seedling rootstocks. Highest success was obtained during January (48.88%) and February (53.33%) and lowest in December (30.24%) and April (34.56%).

Wood apple was successfully propagated at Central Institute for Subtropical Horticulture (CISH), Lucknow by soft wood grafting using cleft method on one year old rootstocks raised in polythene bags having size of 25 X 10 cm. The technique consisted of cutting the rootstock 15 to 20 cm above the ground, making a 3 to 4.5 cm deep cut in the centre, giving two slanting cuts to the proximal end of the 10 to 15 cm long scion shoot without prior defoliation, inserting the scion into the cut made in the rootstock and tying with 200 gauge thick polythene strip. The grafted plants were capped with polythene caps covering the cut and tied portion of the grafted plant, till the plant sprouted well. Polythene bags were filled with soil (sandy loam) : FYM mixture in ratio of 1:1 for raising rootstocks and seeds were sown immediately after extraction from ripe fruits in January. The grafted plants are kept for 20 to 25 days in polyhouse before shifting to shade net house (50%) or a shady place. Care should be taken to select the scion and rootstock having equal thickness and also to remove suckers arising from the rootstock. The graft success of 82 to 96% was achieved during January to March (Shukla *et al.*, 2007).

Planting and after-care

Regular planting of wood apple rarely exists. However, it can be planted at 8 x 8 m spacing in square system. The pits of 1 X 1 X 1 m are dug before the onset of monsoon, preferably during May-June. The *kankar* pan, if any, present in the soil, must be broken completely at the time of digging of pits. Usually, the pits remain exposed to hot sun for at least 15 to 20 days for solarization. Thereafter, the pits are filled up with top soil mixed with well rotten FYM in 1:1 ratio. Besides, 250 g of single super phosphate (SSP) should also be mixed with the mixture. It is always advisable to mix 50 to 100 g of chlorpyriphos dust in the mixture to avoid termite infestation in termite prone areas. The best time for wood apple planting is July to August. The grafted plants, usually in polythene bags, are planted in the centre of the prepared pits after removing polythene bag carefully.

No systematic work has been done on nutritional requirements of wood apple. The fertilizer dose depends on soil fertility and age of the plant. Manure and fertilizer application in sufficient quantities help in improving fruit size and quality. Leaf analysis done at Central Horticultural Experiment Station (CHES now Central Institute for Arid Horticulture (CIAH) Regional Station), Godhra revealed nitrogen (2.016%), phosphorus (2.963 ppm) and potassium (2.325%) contents indicating that the plant removes substantial quantity of these major nutrients from soil which needs to be replenished. Based on experience, a dose 10 kg of well rotten FYM, 100 g nitrogen, 50 g phosphorus and 100 g potassium is applied to one year old plant. This dose is increased every year upto 10 years of tree age. Thereafter, a constant dose of 100 kg FYM, 1 kg nitrogen, 500 g phosphorus and 1 kg potassium is applied. Full dose of FYM, half dose of nitrogen and full doses of phosphorus and potassium are applied during January-February before flowering but after harvesting of fruits. Remaining half dose of nitrogen is given during July–August. The manure and mixture of fertilizers are spread under the canopy of the tree and mixed well in the surface soil. Light irrigation is given immediately after the application of fertilizers.

Wood apple is a crop of dry region and once the plants are established, they hardly need any irrigation. Nevertheless, conservation of rain water in root zone enhances the productivity of this crop. Initially, light irrigations should be given frequently after planting for proper establishment. Later on, irrigations at an interval of 15 days during summer and at monthly intervals during winter improve yield and fruit quality. The tree basin around the tree should be kept free from weeds by regular weeding. If required, glyphosate @ 5 to 7 ml per litre water may be used at times to get rid of perennial weeds. Wood apple trees are allowed to grow freely on central leader with well spaced branches in all directions. Seedling trees attain huge size, while vegetatively propagated trees are dwarf statured. The tree canopy is very compact. Removal of old, diseased, insect pest infested and dead branches and thinning of dense branches ensures quality production and higher yields.

Harvesting and yield

Flowering in wood apple takes place from February to May and ripe fruits are available from October to March. The period of flowering and fruiting is governed by climate and the period of moisture availability. The stage of harvest is determined by the purpose of using the fruit. For chutney, fully developed mature fruits are preferred, however, for squash and jelly, fully ripe fruits are desirable. It takes 210 to 240 days from flowering to maturity. The fruits wrapped individually in newspaper are kept in baskets/cardboard cartons for proper ripening. In some parts of the country, especially in south India, smudging is done to induce ripening. There is possibility to use ethylene for induction of ripening in wood apple. Proper care is required in harvesting of fruits. A minor crack or impact on the rind can cause spoilage during storage. At maturity, the grip of the fruit stalk is loosened and the fruit is detached from the tree without any efforts which is otherwise difficult in unripe fruit. The stem end of the fruit is vulnerable to fungal infection, so it is always desirable to harvest fruits along with small piece (about 1 cm) of fruit stalk.

A vegetatively propagated tree starts bearing after 3 years. However, it takes about 10 years to achieve optimum yield potential. Yield varies considerably but a fully grown up tree can produce about 200 to 250 fruits per annum. Fruits can be graded on the basis of size before marketing. Huge sized wood apple trees of seedling origin growing in forests/road sides have been reported to yield as much as 1000 to 1500 fruits per tree.

References

Anonymous (1956) The Wealth of India. A dictionary of Indian Raw Materials and Industrial products, Vol. IV, CSIR, New Delhi, p. 287.

Bhasin, G.D., Roonwal, M.L. and Singh, B. (1958) *Indian For. Bull. (N.S.)*, **171**: 1-10.

Hebbara, M., Manjunatha, M.V., Patil, S.G., Patil, D.R. (2002) *Karnataka Journal of Agricultural Sciences*, **15**(1): 94-98.

Hiwale, S. S. (2006) In : Advances in Arid Horticulture, Vol. II (Eds. P. L. Saroj and O.P. Awasthi), IBDC, Lucknow, pp. 435-442.

Karale, A.R. (2002) In : Fruits : Tropical and Subtropical, Vol. II (Eds. T. K. Bose, S. K. Mitra and D. Sanyal), Naya Udyog, Kolkata, pp. 731-736.

Kirtikar, K.R. and Basu, B.D. (1935) Indian Medicinal Plants, Vol. I, pp. 496-497.

Maiti, S., Mishra, T.K. (2000) *Ethnobotany*, **12:** 77-80

Roy, P. and Mazumdar, B.C. (1988) *Sci. and Cult.*, **55:** 110-111.

Sappandi, S., Athani, S.I., Patil, P.D., Kanmadi, V.C., Swamy, G.S.K. and Shantappa (2006) Book of Abstracts, Nat. Symp. on Under-utilized Horticultural Crops held at IIHR, Bangalore on June 8-9, 2006, p. 65.

Sheeja, E., Edwin, E. and Smita, G. (2005) *Plant Archives*, **5**(2): 549-552.

Shete, M. B., Karale, A.R., Kaulgud, S.N. and More, T.A. (1999) *Journal of Maharashtra Agricultural Universities,* **24**(2): 176-178.

Shukla, S. K., Singh, A.K. and Singh, M.P. (2006) Book of Abst. Nat.Symp. on Under-utilized Horticultural Crops held at IIHR, Bangalore on June 8-9, 2006, pp.46.

Shukla, S.K., Singh, A.K. and Singh, M.P. (2007) *CISH Newsletter,* **8**(1) : 2

Singh, I.S. (2001) *Indian J. Hort.,* **58** (1&2) : 178-182.

Singh, I. S., Srivastava, A. K., and Singh, V. (1999) *Journal of Applied Horticulture,* **1**(1): 34-37.

Troup, R.S. (1921) In : The Silviculture of Indian trees, Vol. III, Govt of India, pp. 101-103.

Chapter 6 Tamarind

Botanical Name	:	*Tamarindus indica* L.
Regional/ Vernacular Names	:	Imli, amli, anbli, tentul, chintachettu, sintachettu (tree), chintapandu (fruit), puli, amilam, huli, amlam, tentuli, konya
Family	:	Leguminosae

Tamarind is a drought resistant and strong tree which is often planted as wind break. The tree has deep and extensive root system and it can withstand violent typhoons and cyclones. It also tolerates the stiff, salty air of coastal areas. It is an excellent tree for social forestry, agroforestry, wasteland development and dryland horticulture. The tree assumes great significance due to its multiple uses and capacity to withstand adverse soil and climatic conditions. The pulp of the mature, ripe fruit has considerable export value in many parts of the world. It is a fruit with low water content and high levels of proteins, carbohydrates and minerals.

Origin and distribution

Tamarind is indigenous to tropical Africa and some parts of south India. It is widely cultivated in tropical and sub tropical regions of the Africa, Asia and America. It is common in wastelands. It is also found naturalised almost throughout the plains and sub-Himalayan tracts of India, extending to the southern peninsular region. It is grown in home gardens and as an avenue tree.

It is common throughout the warmer parts of India, except where frost occurs *i.e.* sub-Himalayan tracts and north-west Punjab. Rough estimates indicate that about 80000 tonnes of tamarind pulp per annum is produced in India, bulk of which comes from the states of Tamil Nadu, Andhra Pradesh, Maharashtra, Karnataka, Kerala, Orissa, Madhya Pradesh and to a limited extent from Bihar. India is the chief producer and consumer of this fruit in the world. In Maharashtra alone, the area under tamarind is around 13186 ha.

Nutritive, medicinal and other uses

The ripe fruit pulp is composed of 20.9 g moisture, 3.1 g protein, 67.4 g carbohydrates, 5.6 g fibre, 0.1 g fat, 2.9 g total minerals and 230 to 283 k cal energy value per 100 g of edible portion (Anon, 1976). The mineral composition of fruit pulp (on dry weight basis) of common (Brown) and Red type is given in Table 8 (Chauhan *et al.*, 1991).

Table 8 : Mineral composition of two types of tamarind

Type	Per cent					ppm								
	P	K	Ca	Mg	Na	S	Fe	B	Cu	Zn	Al	Sr	Ba	Mo
Brown	0.13	0.81	0.20	0.10	0.12	0.02	54	10	11	18	21	34	3	0.52
Red	0.19	0.82	0.52	0.24	Low	0.03	48	13	14	25	32	47	4	1.02

Tamarind pulp is used widely for flavouring a variety of dishes. Ripe tamarind fruit is said to reduce fever and cure intestinal ailments. Its effectiveness against scurvy is well documented. It is a common ingredient in cardiac and blood sugar reducing medicines. Young seeds contain amber, sweet tasting oil (10 to 15% w/w).

The tamarind is considered to be the sweetest and the sourest of all the fruits due to exceptionally high sugar and acid contents at ripe stage. The total sugars content of tamarind fruit is about 61 per cent of fresh weight (Whiting, 1970). Hasan *et al.* (1978) stated that ripe fruit pulp is fairly rich in sugar content. Total sugars are 18.2 per cent, while free sugars are 17.6 per cent of the pulp.

It is the most acidic of all fruits with a total acidity range varying from 12.2 to 23.8 per cent as tartaric acid (Ulrich, 1970). The unripe (green) fruit is a better source of free tartaric acid than the ripe one. In ripe pulp, almost half of the tartaric acid is in combined form mainly as potassium bitartarate (cream of tartar) and to about 2 per cent of other acids chiefly malic acid and minor amounts of other acids (Kalyankar *et al.*, 1952; Singh, 1973). Pulp has a pH of 3.12 to 3.22 (Savur, 1958). Tamarind pulp is also unique in its high mineral and vitamin contents (Table 9).

Table 9 : Mineral and vitamin contents of tamarind (per 100 g of edible portion)

Minerals	Content	Vitamins	Content
Calcium (mg)	170	β-carotene (μg)	60
Phosphorus (mg)	110	Riboflavin (mg)	0.07
Iron (mg)	17	Niacin (mg)	0.7
Magnesium (mg)	41	Thiamine (mg)	0.44
Copper (mg)	0.20	Vitamin C (mg)	3
Manganese (mg)	0.55		

Several pigments have been identified in different parts of the tamarind. Red pulp variety of tamarind contains the anthocyanin, while a leucoanthocyanidin is present in the brown pulp (common) type. Leaves are very rich in anthoxanthin, while the fruits are poor in this pigment. Xanthophylls are responsible for yellow colour of the flowers.

The sweet acidic pulp is used in the Indian medicine as a refrigerant, carminative, anti-scorbutic and laxative and is also commonly prescribed in febrile diseases and bilious disorders. In India, the pulp is used mainly for culinary purposes; while in other countries, it is processed into nectar fruit punch, juice, sauce, concentrates, glaced and crystalized fruit. It is the fruit base in the preparation of jams, jellies, ice-creams, *sharbats*, beverages including carbonated drinks. Pulp contains 1.27 per cent pectin. Pulp powder is rich in calcium (74 to 143 mg per 100 g), potassium (23.8 to 27.7 mg per 100 g) and copper (1.2 to 3.4 ppm) (Manjunath *et al.*, 1991).

Tamarind seeds contain good amounts of starch, protein and oil. Products from the seeds have many industrial applications. Tamarind kernel powder (TKP) is extensively used for starching of cotton yarns, jute fabrics and woollen. Seed powder also finds use in jelly, fortified bread, biscuits and in the confectionery industry for manufacturing chocolates.

The tender flowers are eaten as vegetable and are considered to be good source of nectar for bees in production of honey having slight acidity. Leaves are regarded as a good fodder with 12 to 15 per cent proteins (Laurie, 1945). Leaves are a good source of calcium. Reddish yellow dye is commercially extracted from leaves. Leaves can be used for green manuring of soil. The bark is given in diarrhoea in lotions and poultices. Bark extracts reduce mosaic virus infection. Timber is invariably used for making agricultural implements, construction work and furniture.

The testa contains colouring agents and water soluble tannins to the extent of 20 to 40 per cent. The use of testa in dyeing and leather tanning industry has been reported. The utilization of testa as a raw material for preparation of plywood and paper adhesive has also been suggested.

The powder obtained by pulverizing deshelled tamarind seed is used in the manufacture of industrial gums and adhesives. Seed powder if added to soil at 0.125 to 1 per cent improves soil structure, water holding capacity and availability of nitrogen (Varade and Badhe, 1969).

The amber (dark brown) coloured fatty acid (6 to 8%) obtained by solvent extraction of deshelled tamarind seed powder is reported to be useful in preparation of paints, varnishes, soap making and also for edible purposes after refining. Oil is also used to give shiny finish to toys. The fatty oil naturally present in kernel powder imparts softness to yarn during weaving and knitting.

Soil and climatic requirements

Tamarind grows successfully in variety of soils ranging from red gravelly loam to degraded wastelands. However, tree thrives best on well-drained deep loamy or alluvial soils. In shallow poor soils, tree remains stunted. It can tolerate saline and alkali soils and can be successfully grown in sodic soils having 49 per cent exchangeable sodium without incorporating any amendments like gypsum and pyrite (Dwivedi *et al.*, 1996).

As regards climatic requirements, tamarind tree has wide adaptability and can be grown in humid to dry hot regions. The tree can thrive well in highly warm climate (48°C) for prolonged period and can tolerate drought. In fact, it is an ideal fruit tree for planting in arid and semi-arid zones of India. The optimum total rainfall requirement per annum is 750 to 1900 mm but can thrive even in regions with a lower rainfall of 500 to 750 mm as in Saurashtra. Young trees are susceptible to frost, while mature trees withstand brief periods of low temperatures (28°F) without much injury. The advent of dry season and moisture stress triggers leaf shedding.

Cultivars

Tamarind is a cross pollinated crop and the existing trees were mostly raised through seedlings. Hence, there is a large genetic diversity within the present population. The east Indian varieties have long pods, with six to twelve seeds, while the west Indian varieties have shorter pods, containing one to four seeds. Seedlings show considerable variation in the size and quality of their fruits. Hence, there is a great scope for survey and selection of superior genotypes from the existing population in the states of Karnataka, Tamil Nadu and Maharashtra. Varieties which yield bigger pods and have high pulp, bold seeds, high total soluble solids and higher tartaric acid content, are preferred in international market. Tamarind varieties have been classified on the basis of various traits. On the basis of pulp colour, there are two distinct types of tamarind :

(i) The yellow or brown pulp type turning dark brown on storage. It is harvested after full maturity.

(ii) The reddish pulp type is locally known as "Raktichinch'. The term 'red type' covers the various shades of pink-pulped fruits. The red type is sweeter than brown (common) type because of lower contents of free acids and is generally harvested when fruits are immature and green. It is mostly preferred for making preserves. A high yielding Red type variety 'Yogeshwari' has been released by Marathwada Agricultural University (MAU), Parbhani, Maharashtra.

On the basis of taste, tamarinds can be classified into two types :

(i) **Sweet type**: The ripe fruits of this type have sweet pulp coupled with less acidity and fruits are mainly used for dessert purpose (Karale, 1998). Makham Waan, Seethong Breed, Manila Sweet are few cultivars found mostly in south-east Asia (Thailand). A sweet type tamarind "No. 263" has been identified by MAU, Parbhani.

(ii) **Sour type**: It is highly acidic in taste and pulp is commercially marketed. The tamarind variety 'Pratisthan' released by Fruit Research Station, Aurangabad, is a sour type and is reported to contain 61 per cent pulp, 12 per cent seed and 27 per cent shell (Anon, 1984). The sour type selections 'Urigam' and 'Cumbum' (good yielder) are popular in Tamil Nadu. A high yielding sour type PKM-1 (Periyakulam-1) was released during 1992 from Horticulture Research Station, Periyakulam, Tamil Nadu (Kumar *et al.*, 1993). This cultivar is preferred for early bearing habit and is claimed to be suitable for high density planting (400 plants per ha as against 100 plants per ha under conventional planting). Patil *et al.* (1997) reported high yielding elite types DT-1 and DT-28 from University of Agricultural Sciences (UAS), Dharwad.

Propagation

Tamarind is mostly propagated by seeds. The seedling trees do not produce true to type plants. The prolonged juvenile phase is also due to seed propagation. The seeds retain viability for several months if kept dry. Normally fully mature seeds of the current season are used for sowing. About 1800 to 2000 seeds weigh one kilogram (Seth *et al.*, 1962). Heavier seeds germinate better and give rise to more vigorous seedlings as compared to those from lighter seeds.

Multiplication of elite/promising types of tamarind by vegetative means is a must for producing true to type plants. It can be propagated by air layering, grafting, budding, cutting and stooling. Soft wood grafting has become the most common and successful method for its vegetative propagation.

Air layering is also an important method of tamarind propagation. Shoots selected for air layering should be 40 to 45 cm long and 1 cm in diameter and preferably from previous year's growth. A ring of bark, about 3 cm long, is removed and covered with sphagnum moss, previously soaked in water, and wrapped with polythene film. Both ends of the film are carefully tied and left for rooting. Bhore (1986) reported that air layers made in brown pulped (sour) type gave 60 per cent success during September by applying 4000 ppm IBA in lanolin followed by IBA 5000 ppm (54% success). In case of red pulp type, air layers with IBA 4000 ppm gave 50% success followed by IBA 5000 ppm (40% success). Navaneetha *et al.* (1991) reported that air layers made in July with 1000 ppm IBA resulted in higher percentage of rooting (75%) and survival (24.38%) followed by 5000 ppm of IBA in the same month. Lokhande (1978) reported that air layered plants have comparatively weaker root system than seedling plants and the field success of air layered plants need utmost care. Gowda (1983) observed that girdling and etiolation with IBA or IBA+NAA increased the number and length of air layered roots. Ganapathi (1989) reported that rooting percentage in air layers of tamarind increased with IBA at 2500 ppm. This treatment also produced more roots per shoot (Ganapathi *et al.*, 1998) from a 20-year old tree. Patil *et al.* (1997) obtained 75 per cent success by treatment with 2000 ppm IBA during June and July months followed by August layering (60%) with 1000 ppm IBA+NAA treatment. Karale (2001) reported that mid-August was the best season with 82 per cent success followed by mid-July layering.

The technique of mound layering consists of cutting down the selected tree to the ground level. Within two or three months, many buds start sprouting from the ground level. These are covered with sand and soil up to 15 to 20 cm and after 30 to 40 days the shoots which turn from green to brown are ringed and treated with IBA and again covered with sand. The ringed shoots produce roots in 30 to 40 days. The sand around the stump is removed and the shoots are separated for planting. The number of layers that can be obtained by mound layering is low. In mound layering studies, highest percentage of rooting (80%) was recorded in 2000 ppm IBA treated shoots followed by 70% rooting with 2500 ppm IBA (Gowda, 1997).

Approach grafting is one of the common methods of vegetative propagation for tamarind. The rootstocks are raised in containers. Attempts were made by Bhore (1986) to propagate tamarind by inarching during November on two-year old seedling stock, having stem thickness of about 0.7 to 0.8 cm and obtained 95 per cent success. Swaminath and Ravindran (1989) recorded more than 60 per cent success of approach grafting. Karale (2001) reported that approach grafting gave success of 52 to 88 per cent during mid-June to mid-September. Gowda (1997) reported 80 per cent success in green wood cleft grafting done in May followed by April and June. Hiremath *et al.* (2006) recorded cent per cent rooting in air layers in SMG-13 and NTI-75 genotypes, while 85 per cent success was achieved with soft wood grafting in NTI-19.

Purshotam and Rao (1990) reported 49 per cent success with veneer grafting in April. They further reported that 10 cm long scion sticks were defoliated before grafting. Veneer grafting was done on 6-month old seedling rootstocks at 8 cm height and the stock was cut above the graft union immediately after the operation.

In recent years, soft wood grafting has been standardized and it is commercially employed in Maharashtra and other parts of country. This is by far the best method of tamarind propagation. Purshotam and Rao (1990) reported that percentage graft take in soft wood grafting was better (68%) and number of sprouts were also comparatively higher (3.22) than veneer grafting but average sprout length was marginally less (15.95 vs. 16.69 cm). Soft wood grafting on 8 to 10 month old seedling stock (22 to 32 cm height and 0.3 to 0.4 cm diameter) in the month of March-April gave over 75 per cent success (Kulwal *et al.*, 1997). Salve (1994) noted 65 per cent survival by wedge grafting on 10th April whereas grafting done on 10th March recorded the highest growth rate. The scion sticks having 10 cm length proved to be superior in respect of final survival (72%) and growth rate. Shinde *et al.* (1996) reported that soft wood grafting in March or in the first week in April (after harvesting and before flowering) resulted in highest grafting success (58.4 to 70%). Wedge method during 2nd fortnight of January to 1st fortnight of February gave 75 to 100 per cent success (Gojamgunde *et al.*, 1997). Soft wood grafting in August-September was found suitable under shade, which gave up to 60 per cent success (Naik *et al.*, 1997). Bharad *et al.* (1999) recorded the highest percentage of graft success and survival by grafting in March-April.

Gayake (1992) reported that soft wood grafting done on 5th February at Parbhani gave maximum success of 33.05 per cent. It was followed by 15th (28.33%) and 25th March (20.46%). Under similar climatic conditions, Satisha *et al.* (1997) reported that soft wood graft success was associated with rootstock age. At 60 days after grafting, success was significantly higher for 6 and 9 month old rootstock (69 to 72%) than for the others (3, 12, 15 or 18 month old rootstocks). Karale (2001) reported success ranging from 6 to 93 per cent during mid-March to mid-August through soft wood grafting in tamarind. Substantial difference was observed with regard to days taken for sprouting. Grafting done in mid-May required considerably shorter duration (14 to 25 days).

Budding can also be employed for propagating tamarind. Prominent buds are taken from mature shoots of previous year's growth. The seedlings used as rootstock are prepared by removing the side growth up to 20 to 25 cm from the base. The budding may be done either by shield or T or by the ring and patch method. Pathak *et al.* (1991) obtained 96 per cent take in patch budding and 94 per cent success in modified ring budding on 9-month old rootstocks. Karale (2001) reported maximum success of 63 per cent through patch budding in mid-July under Rahuri conditions.

Planting and after-care

The tamarind tree attains a huge size, it should be planted 10 to 12 m apart in square system in the field or 10 m apart as an avenue tree or for farm boundary plantation. After planting, seedling/grafts need to be protected from stray cattle. Regular watering and weed removal from the basins of young plants help in proper establishment of plants. Mulching with dry grass should be done in the basin of the plant to conserve the soil moisture.

Tamarind trees are rarely manured. Scientific information is not available on this aspect. But for sustained production, it is necessary to give manures and fertilizers regularly. Karale (2006) suggested the following schedule for tamarind (Table 10).

Table 10 : Nutritional requirements of tamarind

Age of tree (Yrs.)	FYM (kg)	Dose of Fertilizers (g/tree)		
		Ammonium sulphate	Urea	Mixed fertilizer
Non-bearing trees				
1	5	50	25	25
2-3	10	100	50	25-50
4-5	15	200	100	50-75
6-7	25	250	100	100
Bearing trees				
8-15	30	250-500	100-200	
16-24	40	500-1500	200-500	250-500
25 and above	50	1500-2500	500-1000	500-750

FYM and mixed fertilizers should be applied in June-July and ammonium sulphate or urea in two split doses in mid-August and mid-September. Manures and fertilizers should be applied when there is sufficient moisture in the soil under the tree canopy leaving 0.5 to 1.0 m around the tree trunk. After application, fertilizers are mixed with the soil. Micronutrients particularly iron may be required for trees in alkaline soils.

Though tamarind can tolerate drought conditions, it responds well to irrigation. The soil water balance influences the triggering of leaf shedding, sprouting of buds, fruit set and fruit retention. Young plants (pre-bearing age) should be irrigated frequently during dry season and long breaks in monsoon in winter and summer at an interval of 7 to 10 days. Irrigation should be given within the periphery of the canopy. Once the plant has established, it hardly needs any irrigation under humid tropics. However, under arid conditions, run off collection techniques should be adopted to facilitate maximum water harvest in root zone.

Training is essential for tamarind during initial 2 to 3 years. This is done with a view to providing a good framework for the future so that the branches are spread properly and these do not break with the crop load at the bearing stage. The branches are not encouraged too low on the trunk or too high from the ground level. In this way 1.5 to 2.0 m trunk from ground level is maintained and three to five well spaced branches are allowed to develop into the main scaffold structure of the tree. The tamarind being an evergreen plant, hardly needs any pruning. It is only required to remove dead or damaged wood.

Old and inferior seedling trees can be converted into productive ones of the desired superior type by dehorning the main limbs of the tree, allowing new soft shoots to produce and using soft wood grafting technique (Karale, 1998).

Harvesting and yield

Tamarind flowers from April to July in southern, northern and western India. In the eastern parts of the country, new leaves appear in May and flowering continues up to September. The flowers are borne in small terminal drooping racemes on the current season's growth. It is a protogynous and highly cross pollinated crop. Pollination is carried out by flies, bees and red ants. The fruit set is the most potent factor, which determines the yield. Though the flowering is profuse in tamarind, the fruit set under natural conditions is low ranging from 3 to 5 per cent, which can be increased to about 50 to 55 per cent under artificial cross pollination. In India, fruits generally mature in February to March. The pods should be allowed to ripen on the tree until the epicarp (outer shell) gets dry and hardened and separated from the pulp without adherence. The ripe pods are collected from the trees by shaking the branches. The ripened pods, if not picked, may remain hanging for almost one year after flowering. The dry ripe pods are gathered and shells are removed manually using a wooden stick. The pulp is separated from the seeds and rags (fibres) and dried in the sun to reduce its moisture level. When harvested, the fruits are almost semidried and do not spoil due to high acidity. The pulp is packed in gunny bags and stored in dry place.

A seedling tree takes longer time to come into fruiting ranging from 7 to 10 years whereas grafted plants are precocious and come into bearing in about 4 to 5 years. As tree grows in size and age, productivity increases. A full grown tree produces about 200 kg of fruits per annum. A well managed tree may bear 300 to 750 kg of ripe pods per year. Kumar *et al.* (1993) reported that clonally propagated selection PKM-1 yields 260 kg ripe fruits as against 165 kg per tree in local cultivar.

The usual practice is to auction the tree/orchard to the contractors at flowering stage and the traders collect/purchase unseeded pulp for stocking in cold storages for sale in the subsequent year. The crop from

Karnataka and Andhra Pradesh arrives in the market in January, from Maharashtra it arrives in February, from Madhya Pradesh and Uttar Pradesh by the end of February. The major markets for tamarind are Jagdalpur and Indore in Madhya Pradesh, Ranchi in Bihar, Chittoor in Andhra Pradesh, Belgaon and Tumkur in Karnataka, Kolkata and in Maharashtra, Vashi (Mumbai) for un-seeded pulp and Barsi (District Solapur) for seed.

References

Anonymous (1976) The Wealth of India. Raw Materials, Vol XI, CSIR, New Delhi, pp. 114-122.

Anonymous (1984) Cash Crop Science Sub-committee Report. MAU, Parbhani (Maharashtra).

Bharad, S.G., Gholap, S.V., Dod, V.N. and Bople, S.R. (1999) *J. Applied Hort.*, **1** (1) : 57-58.

Bhore, D.P. (1986) *Maharashtra J. Hort.*, **3**(1) : 55-58.

Chauhan, K.S., Pundir, J.P.S. and Singh, S. (1991) *Haryana J. Hort. Sci.*, **20:** 210-213.

Dwivedi, S.K., Ali, W. and Pathak, R.K (1996) *Annals Agric. Res.*, **17**(4) : 447-449.

Ganapathi, T. (1989) *Mysore J. Agric. Sci.*, **21** : 435.

Ganpathi, T., Samiullah, R. and Naik, K.R. (1998) *Karnataka J. Agril. Science*, **11**(1) : 266-267.

Gayake, H.P. (1992) M.Sc. (Ag.) Thesis, MAU, Parbhani.

Gojamgunde, M.D., Chakrawar, V.R., Kadam, B.A. and Shinde, B.N. (1997). Abstr. Proc. Nat. Symp. on *Tamarindus indica* L, June 27-28 , Tirupathi, p.14.

Gowda, N. (1983) M.Sc. (Hort.) Thesis, UAS, Bangalore.

Gowda, N.V. (1997) Abstr. Proc. Nat. Symp. on *Tamarindus indica* L. June 27-28, Tirupathi, p. 15.

Hasan, S.K., Salam, A. and Wahid, M.A. (1978) *Pakistan J. Sci.*, **30** (1-6): 71-73.

Hiremath, S.M., Gowda, D.S.M. and Budihal, R.A. (2006) Book of Abstracts, Nat. Symp. on Under-utilized Horticultural Crops held at IIHR, Bangalore on June 8-9, 2006, p.74.

Kalyankar, G.D., Krishaswamy, P.R. and Sreenivasaya, M. (1952) *Curr. Sci.*, **21:** 210-222.

Karale, A.R. (1998) Status Report on "Improvement of Tamarind by selection" presented at Research Workers Group meeting of AICRP on Arid Zone Fruits held at CCSHAU, Hissar, Oct. 14-16, 1998.

Karale, A.R. (2001) Final Report of ad-hoc project on tamarind submitted to ICAR, New Delhi, (Period 1.1.1998 to 31.12.2000).

Karale, A.R. (2006) In: Advances in Arid Horticulture, Vol. II (Eds. P.L.Saroj, and O.P.Awasthi), IBDC, Lucknow, pp. 405-433

Kulwal, L.V., Tayade, G.S., Choudhari, M.H. and Verma, R.M. (1997) Abstr. Proc. Nat. Symp. on *Tamarindus indica* L. June 27-28, Tirupathi, p. 9.

Kumar, N., Arunugam, R., Kedar, M.M. and Irulappan, I. (1993) Proc. Golden Jubilee Symp. Hort. Soc., India, Bangalore, Abstr., p. 26.

Laurie, M.V. (1945) *Indian Forester Leaflet 82*, FRI and College. Dehradun, p. 17.

Lokhande, R.V. (1978) MSc. (Ag.) Thesis, MAU, Parbhani.

Manjunath, M.N., Sattigeri, V.D., Rama Rao, S.N., Usha Rani, M. and Nagarja, K.V. (1991) *Indian Fd. Packer*, **45**(3) : 39-42.

Naik, K.R., Sulikeri, G.S., Patil, V.S. and Basavaraj, N. (1997) Abstr. Proc. Nat. Symp. on *Tamarindus indica* L., June 27-28, Tirupathi, p.10.

Navaneetha, N., Palaniswamy, K.P., Khader, Md. Abdul and Kumar, N. (1991) *South Indian Hort.*, **39**(2) : 102-105.

Pathak, R.K., Ojha, C.M. and Dwivedi, R. (1991) *Indian Hort.*, **36**(2) : 17.

Patil, V.S., Madalageri, M.B., Sulikeri, G.S. and Naik, R. (1997) Abstr. Proc. Nat. Symp. on *Tamarindus indica* L., June 27-28, Tirupathi, p. 9.

Purshotam, K. and Narsimha Rao, S.B.S. (1990) *South Indian Hort.*, **38**(4) : 225

Salve, P.D. (1994). M.Sc. (Ag.) Thesis, MPKV, Rahuri.

Satisha, J., Melanta, K.R. and Venkatesha, J. (1997) *Curr. Res.*, **26** (6-7) : 93-94.

Savur, G.R. (1958) *Chem. Abstr.*, **52** : 17413-a.

Seth, S.K., Raizado, M.B. and Waheed Khan, M.A. (1962) In : Trees for Vana Mahotsava, F.R.I. and Colleges, Dehradun, p. 130.

Shinde, N.N., Ingle, C.N. and Shirurkar, P.D. (1996) *J. Applied Hort.*, **2**(1-2) : 139-142.

Singh, P.P. (1973) *Qual. Plant Mater. Veg.*, **22** : 335-347.

Swaminath, M.H. and Ravindran, D. S. (1989) *My Forest*, **25** (4) : 357-360.

Ulrich, R. (1970) In : The Biochemistry of Fruits and their Products. Vol. 1, (Ed. A.C. Hulme), Academic Press, New York.

Varade, P.A. and Badhe, N.N. (1969) Nagpur Agric. Coll. Mag., **41** : 47-51.

Whiting, G.C. (1970) In : The Biochemistry of Fruits and their Products. Vol. 1. (Ed. A.C. Hulme), Academic Press, New York.

Chapter 7 Jackfruit

Botanical Name	:	*Artocarpus heterophyllus* Lam. syn. *Artocarpus integra* (Thunb.) Merril; *Artocarpus integrifolia*
Regional/ Vernacular Names	:	Kathal, panasa, kanthal, phanas, pilapalam, halasu, chakka
Family	:	Moraceae

The jackfruit is hardly regarded as a commercial fruit crop although it is widely cultivated in southern Asia, the East Indies and other warm areas of both hemispheres (Morton, 1965). It is also found growing in Myanmar, Malaysia and Brazil and is very popular in eastern and southern parts of India. Jackfruit has the largest fruit size among the edible fruits.

Origin and distribution

The jackfruit is native to India and is now widely cultivated throughout the tropical low land in both the hemispheres. In south-east Asia, it is widely grown commercially and in the home gardens. The major jackfruit producing countries are Bangladesh, India, Myanmar, Indonesia, Sri Lanka and Malaysia. The large size of the fruit, weighing 30 kg or more, unusual appearance and strong aroma contribute to the jackfruit's slow acceptance (Campbell *et al.*, 1998).

Nutritive, medicinal and other uses

Jackfruit appears in the market almost throughout the year as a popular vegetable. It enjoys a high demand and premium price. Ripe fruits have high nutritive value and are available during June-July. Fruit has 21 to 51% pulp, 16 to 21% TSS and 0.08 to 0.12% acidity (Singh, 2001). Zaghlol *et al.* (1983) noted glucose (10.94%), fructose (1.42%), xylose (0.18%), rhamnose (0.44%), arabinose (1.51%), galactose (2.47%), and three other unknown sugars (13.4%) in ripe jackfruits. Jackfruit is an important source of pectin (Krishnamurthi and Giri, 1949) and contains about 1.9 per cent protein on fresh weight basis. Hossain *et al.* (1979) stated that the fruit contained 2.64 to 11.77 mg of ascorbic acid and 250 to 1740 mg carotene per 100 g of edible portion. The jackfruit latex is reported to possess bacteriolytic activity. The composition of edible portion is detailed in Table 11 (Anon, 1979).

Table 11 : Composition of jackfruit (per 100 g of edible portion)

Constituent	Tender	Ripe	Seed
Moisture (%)	84	77.2	64.5
Carbohydrates (g)	9.4	18.9	25.8
Protein (g)	2.6	1.9	6.6
Fat (g)	0.3	0.1	0.4
Fibre (g)	-	1.1	-
Total minerals (g)	0.9	0.8	1.2
Calcium (mg)	50	20	21
Phosphorus (mg)	97	30	28
Iron (mg)	1.5	500	-
Potassium (mg)	246	-	-
Vitamin A (IU)	0	540	17
Thiamine (mg)	0.25	30	-
Riboflavin (mg)	0.11	-	-
Vitamin C (mg)	11	-	-

Kumar *et al.* (1998) also noticed that seeds of jackfruit are good source of carbohydrates (26.83 to 28.01%), protein (6.25 to 6.75%) and minerals (1.16 to 1.27%). Jackfruits are eaten unripe at 25 to 50% full size as a vegetable or ripe as a fruit. When used as vegetable, it is peeled, sliced and boiled and then seasoned or mixed with other food. Ripe fruit has high nutritive value. In India, it is called as the poor man's fruit. The ripe fruit also significantly contributes to the nutrition of low income families as a source of vitamins, minerals and calories. It contains fairly good amount of β-carotene, a precursor to vitamin A. Jackfruit is an important source of pectin and contains about 1.9 to 2.2 per cent protein on fresh weight basis. Fructose, glucose and sucrose are the major sugars in all parts of the fruit, except in the outer spiny rind, which is devoid of glucose. Capric, myristic, lauric, palmitic, oleic, stearic, linoleic and arachidic acids are the major fatty acids, with varying proportions in different parts of the fruit (Chowdhury *et al.*, 1997).

Several value added products can be processed from both green and ripe fruits (Singh, 2001). Green mature fruits are used for making the pickles. Ripe fruits may be processed into jam, squash, ready-to-serve beverage, chutney and *papad*. Fruit at early stage of ripening may be utilized for making preserve and candy.

The seeds are boiled or roasted and used in many culinary preparations. The seeds are very rich in carbohydrates and protein. Incorporation of jack-seeds to deep fat fried products revealed that it brings down fat absorption to a remarkable extent with affecting sensory attributes of savoury product (Rajarajeshwari and Prakash, 1999). The leaves and skin of the fruit are excellent cattle feed. The latex from the bark contains resins. Sometimes this is used to plug holes in earthen vats and buckets. The skin and core of the fruits, could be used commercially for pectin extraction.

Soil and climatic requirements

The jackfruit can grow on a wide range of soils, though it prefers a rich, well drained, deep alluvial soil. Rise in water table or flood or water logging severely damages the trees and may lead to death. Jackfruit trees can be grown on open-textured light soil or laterite soil provided nutrients are available.

Warm humid plains are suitable for jackfruit. It flourishes in humid hill slopes also up to an elevation of 1500 m. Quality of fruits deteriorates in higher altitudes. It also grows well in arid, semi-arid and warmer plains of south India. Cold weather and frost are harmful.

Cultivars

Jackfruit is cross-pollinated and mostly seed propagated resulting in innumerable types or forms considering the fruit characteristics. The types differ widely among themselves in bearing, size, shape, quality and period of maturity. Wide variations, *i.e.* sweetness, acidity, flavour and taste are also observed. Such variations among clone offer scope for improvement of this crop by clonal selection method. No systematic breeding programmes have been undertaken, but many local varieties exist. In India and Bangladesh, selection programmes have started and several clones have been identified and are now being evaluated for release. Ripe jackfruits are divided into two types, based on edible pulp. The first type has thin, fibrous, soft edible flesh, acid to very sweet with a strong aroma. The other type has thick, firm to crisp flesh with less aroma. There are many varieties of both types. Some are more suited to canning than other fresh fruit varieties.

In India, mainly two types, soft flesh and firm flesh type are available. Some types with local names like 'Gulabi' (rose-scented), 'Champa' (flavour like that of *Michelia* sp.); 'Hazari' (bearing large number of fruits), 'Rudrakshki' (roundish fruits) are also available.

A genotype (No. 7) was selected from Uttar Pradesh, with moderate yields, large fruits weighing 15 kg or more and bulbs around 20 g, small cylinders, high pulp/cylinder ratio and average fruit quality (Kumar and Singh, 1996). In West Bengal, about 35 types were identified as superior clones. Wide variability in yield (14 to 1200 fruits per tree), fruit weight (2.0 to 16.0 kg), fruit shape (oblong, roundish, conical); peel colour, pulp colour, number of flakes (30 to 380 per fruit), number of seeds (30 to 365 per fruit), texture of pulp, pulp weight, total soluble solids (15 to 30%), total sugars (13 to 27%) and acidity (0.10 to 0.33%) were observed among the genotypes. The growth, bearing (twice a year in some types) and maturity (June-September) of fruits also exhibit variations among the genotypes (Mitra and Maity, 2000). Salvi *et al.* (2006) evaluated seven genotypes of jackfruit at Regional Fruit Research Station, Vengurle and a promising genotype KKV/VGL/JF 9/93 has been released as Konkan Prolific. It is heavy bearer, precocious with fruits of excellent quality and medium size (8 to 10 kg per fruit) and is suitable for fresh consumption, processing and marketability.

Propagation

The most common method of propagating jackfruit is by seed. Generally, 4 to 5 seeds are sown *in situ* so that the tap root can grow undisturbed. However, seedlings can be raised in pots or polybags. After one or two years, the seedlings are ready for planting in the field. Germination of seeds deteriorates on storage. Seventy per cent of the seeds germinate within 15 days after extraction. But, after 30 day storage, the germination declined to 40 per cent. Larger seeds germinate better. Soaking of seeds in water for 24 hours was found to improve the germination. Krishnasamy (1990) suggested that the critical seed moisture content in a cultivar was 39%, at this moisture content level 93% of seed germinated, approximately. Freshly extracted seeds have a moisture content of 62% and they give 100% germination.

A number of vegetative propagation methods have been attempted by different workers to reproduce true to type plants. Soft wood grafting was tried in jackfruit using mist chamber and green house structures (Patil *et al.*, 1998). Grafting of scion shoots taken from one season old terminal shoots of jackfruit on 2 month old seedling rootstocks during second fortnight of June to July gave 60 per cent graft success. The grafts were kept in mist chamber till sprouting, before transferring them to green house for further growth. Mokashi *et al.*

(2006) obtained 51.9 per cent success with soft wood grafting of jackfruit on 7 month old rootstocks at University of Agricultural Sciences (UAS), Dharwad. Maiti and Sen (1999) observed maximum success (60%) by ringing and blanching of scions for 3 days before grafting. Swamy *et al.* (1998) recorded 40 to 50% success with soft wood grafting under open conditions.

Epicotyl grafting with mature, plump, terminal scion shoot on germinating jackfruit seedling of about 8 to 10 days by wedge method during April-May gave 50 to 90 per cent success and the successful epicotyl grafts attained saleable size within a year (Gunjate *et al.*, 1982). Jose and Velsalakumari (1991) reported highest percentage of sprouting (83.30%) and survival (61.67%) for grafting done during June. Epicotyl grafting with 3 to 4 month old scions and 5 day old rootstocks resulted in the highest percentage sprouting (29.50%) and the highest percentage survival (21.67%). Moist conditions increased the percentage of success in both kinds of grafts.

Mukherjee and Chatterjee (1979) reported 84 per cent success in rooting of cuttings by etiolation and forcing of shoots followed by treatment with 5000 ppm IBA, and then keeping under a mist. No rooting was obtained without dipping in IBA. Survival of cuttings was 50 to 70 per cent. Dhua *et al.* (1983) recorded maximum rooting success (90%) from shoots which were etiolated and ringed for 30 days and then treated with IBA at 3000 ppm in combination with ferulic acid at 2000 ppm. Pre-conditioning (especially etiolation) or stock plant treatment with ethephon before taking cutting and treating the cuttings with IBA (5000 ppm) and p-hydroxybenzoic acid (200 ppm) at the time of planting was found effective in inducing rooting of semi-hard wood cuttings (Mitra and Bose, 1996).

Sen and Bose (1959) reported 100 per cent success in rooting of air layers. Treatment with IBA markedly improved the root formation. Pre-treatments of shoots like girdling, etiolation and treatment with IBA and NAA were found more effective in rooting of layers when applied in combination rather than when applied individually (Lingarajappa, 1982). Nazeem *et al.* (1984) recorded 81.29 per cent rooting in air layers of one-year old shoots of bearing plants when coconut pith was used as rooting medium. Datta (1988) recorded 100 per cent rooting by pre-treatment of stock plants with ethrel (250 ppm) or cycocel (1000 ppm) and application of IBA at 3000 or 5000 ppm. Alila *et al.* (2000) concluded that the treatment with IBA (5000 ppm) could be used for adventitious root formation in air layering of jackfruit in combination with etiolation for a period of 30 days. Stool layering has also been tried in jackfruit. Alila *et al.* (1997) reported that medium sized stool resulted into more number of bud emergence. IBA at all (500, 1500, 3000 and 4500 ppm) but preferably lower concentration proved better than NAA (250, 500, 750 and 1000 ppm) in rooting of layered shoots.

Teaotia *et al.* (1963) reported 100 per cent success by patch budding. Patch, chip and 'T' methods of budding were tried during February to September by Singh *et al.* (1982). The maximum success (90%) was obtained following patch method in June and 'T' budding was found least successful.

Planting and after-care

The square system is commonly followed for planting. A spacing of 10 x 10 m accommodating 100 plants per ha is sufficient for this fruit crop. Higher density of planting can be practised in lighter and poorer soils. The spacing can be increased to 12 X 12 m in fertile soil to avoid crowding of trees in later stages.

For planting jackfruit, pits of 1 X 1 X 1 m size are dug at least 15 to 20 days before planting. After exposure of the soil to the sun for sufficient period, 30 to 40 kg well-rotted farm yard manure and 500 g single super phosphate are mixed with the top soil of each pit followed by filling of pits. Chlorpyriphos dust @ 50 to 100 g per pit may be mixed with the mixture to avoid the termite attack. After planting, the soil is pressed firmly to avoid water logging in pits during rainy season. The best time for planting grafts or seedlings is July-August in north India. Prolonged dry weather after planting may lead to death of plants. The tap root should not be disturbed while planting to avoid damage to plants.

The young plants are frequently browsed by stray goats and cattle unless adequately guarded initially for about two years. Hand watering of young plants during initial years is necessary for assured survival and good growth of plants. In colder regions, protection against frost at least during first few years is a must. Cleaning of basins by regular weeding/interculture and mulching in tree basins are good for proper maintenance of orchard.

The jackfruit is not irrigated normally. The trees are sensitive to drought. Irrigation during dry periods is considered essential in arid regions for normal growth. In order to economise use of water, ring system may be adopted for irrigation. For young orchards, hand watering is necessary during first 2 to 3 years. The frequency of irrigation depends on the soil moisture condition.

Jackfruit trees need nutrition for regular and good cropping. Application of farm yard manure (37 kg per plant), 750 g ammonium sulphate, 625 g single super phosphate and 300 g muriate of potash per plant per year in the second and third year of planting and in the fourth year an increased dose of 938 g ammonium sulphate, 781 g single super phosphate and 375 g of muriate of potash were found to increase growth, flowering and fruiting (Mukherjee *et al.,* 1983).

For quick growth of trees, manures and fertilizers may be added twice a year, before and after monsoon. It is advisable to apply 80 kg farm yard manure per tree per year. In addition, according to the nutrient status of the soil and growth of the trees, chemical fertilizers should be applied for desired results.

The fertilizers and manures should be spread in the basin and thoroughly mixed in the soil. The fertilizers should be applied in an area under the tree canopy leaving 1 to 1.5 m from the trunk.

Harvesting and yield

The tree starts bearing from the seventh or eighth year after planting. The flowers generally start appearing in December and continue up to March and the fruit ripens in summer. Tender jackfruits are harvested for use as vegetable during early spring and summer until the seeds harden. The fruit matures towards the end of summer or start of monsoon in June-July. Period of fruit development is February to June. The optimum stage of harvest maturity has been reported to be 90 to 110 days after the appearance of the spike. Harvesting is done by cutting off the footstalks carrying the fruits.

The tree reaches its peak bearing stage within fifteen to sixteen years after planting. Normally, a tree bears a few to about 250 fruits annually at this stage. The weight of fruits varies widely depending on the type. Individual jackfruit may weigh from about 1 kg to about 20 kg.

References

Alila, P., Chiesotsu, S., Kar, P.L. and Sanyal, D. (2000) *The Hort. J.,* **13** : 9-13.

Alila, P, Sanyal, D., Kar, P.L. (1997) *Crop Research Hissar,* **14**(3): 433-436.

Anonymous (1979) Extension Bulletin No. 22, IIHR, Bangalore.

Campbell, R.J., El-Sawa, S.F. and Eck, R. (1998) *Fairchild Hort. Series,* Vol.2, p. 23.

Chowdhury, F.A., Raman, M.A. and Mian. A.J. (1997) *Food Chemistry*, **60:** 25-28.

Datta, P. (1988) MSc. (Ag.) Thesis BCKV, Kalyani.

Dhua, R.S., Sen, S.K. and Bose, T.K. (1983) *Punjab Hort. J.*, **23** : 84-91.

Gunjate, R.T., Kokekar, D.T. and Salvi, M.J. (1982) Proc. Nat. Sem. on Plant Propagation, Kolkata, p.87.

Hossain, M., Haque, A and Hossain, M. (1979) *Bangladesh J Agric.*, **4:** 9-12.

Jose, M. and Velsalakumari, P.K. (991) *South Indian Hort.*, **39** : 264-267.

Kabir, S. (1998) *J. Immunologica Methods*, **212**: 193-211.

Krishnamurthi, C.R. and Giri, K.V. (1949) *Proc. Indian Acad. Sci.*, **298** : 155-167.

Krishnasamy, V. (1990) *South Indian Hort.*, **38** : 218-219.

Kumar, S., Singh, A.B., Abidi, A.B., Upadhyay, R.G. and Singh,A. (1988) *J.Fd. Sci. Technol.*, **25**(5): 308-309.

Kumar, S. and Singh, I.S (1996) *Recent Hort.*, **3** : 55-59.

Lingarajappa, V.G. (1982) M.Sc. (Ag.) Thesis, UAS, Bangalore.

Maiti, C.S. and Sen, S.K. (1999) *The Hort J.*, **12** . 97-99.

Mitra, S.K. and Bose, T.K. (1996) *Scientific Horticulture*, **5** : 1-7.

Mitra, S.K. and Maity, C.S. (2000) Symp. Booklet, International Symp. on Tropical and Sub- tropical Fruits, Cairns, Australia, p. 45.

Mokashi, A.N., Hegde, R.V. and Aralikatti, G. (2006) Book of Abstracts, Nat. Symp. on Under-utilized Horticultural Crops held at IIHR, Bangalore on June 8-9, 2006, p.67.

Morton, J. (1965) *Proc. Fla. Sta. Hort. Sci.*, **78** : 336-344.

Mukherjee, S.K and Chatterjee, B.K. (1979) *Scientia Hort.*, **10** : 295-300.

Mukherjee, S.K, Rao, D.P., Mukherjee, S., Ray, M. and Mohanty, B.B. (1983) *Indian J. Hort.*, **40** : 150-154.

Nazeem, P.A., Gopikumar, K. and Kumaran, K. (1984) *Agric. Res. J. Kerala*, **22:** 149-154.

Patil, V.S., Sulikeri, G.S., Mannikeri, I.M. and Patil, S.S. (1998) Souvenir-Nat. Sem. on New Horizons in Production and Post Harvest Management of Tropical and Subptropical Fruits held at IARI, New Delhi on Dec. 8-9, 1998, p.6.

Rajarajeshwari, H. and Prakash, J. (1999) *The Indian J. Nutrition and Diet*, **36:** 312-319.

Salvi, B.R., Malshe, K.V., Bendale, V.W., Kshirsagar, P.J. and Gurav, S.S. (2006) Book of Abstracts, Nat. Symp. on Under-utilized Horticultural Crops held at IIHR, Bangalore on June 8-9, 2006, p. 32.

Sen, P.K. and Bose. I.K. (1959) *Indian Agric.*, **3:** 43-47.

Singh, I.S. (2001) *Indian J. Hort.*, **58** (1& 2) : 178-182.

Singh, U.R, Pandey, I.C. and Upadhyaya, N.P. (1982) *Punjab Hort. J.*, **22** : 103-105.

Swamy, G.S.K., Melanta, K.R., Athani, S.I. and Patil, P.B. (1998) Souvenir-Nat. Sem. on New Horizons in Production and Post Harvest Management of Tropical and Subptropical Fruits held at IARI, New Delhi on Dec. 8-9, 1998, pp. 8-9

Teaotia, S.S., Dayal, K. and Asthana, M. P. (1963) *Sci. and Cult.*, **29** : 46-47.

Zaghlol, H.A., Elohamy, L.M., Moustafa, S.M.L. and El-Olemy, M.M. (1983) *Sci. Pharma,* **51** : 391-396.

Chapter 8 Lasora

Botanical Name	:	*Cordia myxa* Roxb. syn. *Cordia dichotoma* Forst
Regional/ Vernacular Names	:	Lasoda, labhera, lehsua, Indian cherry, cherry of the desert, sebestan, gonda, leshora, bahuvaraka, bahubara, shelvant, bargund, chinna nakkeru, naruvili, chikkachalle, viri, cheruviri
Family	:	Boraginaceae

Lasora is gaining popularity in relatively dry and rainfed areas of the country due to its higher productivity, suitability to adverse soil and climatic conditions, high processing value and good market. It gives high economic returns even with little care and inputs. Lasora tree is also valued for its cool and comfortable shade during hot summer. The trees can be planted around the orchards as windbreak. It is a fairly fast growing tree.

Origin and distribution

Lasora is native to India. It is found growing abundantly in Rajasthan, Punjab, Haryana, Uttar Pradesh, Madhya Pradesh, Chattisgarh, Gujarat, Maharashtra, Assam, Meghalaya, etc. In Uttar Pradesh, it is found throughout the state, but more commonly in Jhansi, Mahoba, Hamirpur, Chitrakut, Etawah, Agra, Mathura, Mainpuri districts and adjoining areas. There is good demand for lasora fruits in Rajasthan, Gujarat and parts of Uttar Pradesh and Haryana bordering Rajasthan.

Nutritive, medicinal and other uses

The fruit contains about 70 per cent mucilaginous pulp, average moisture (70%), total soluble solids (10%) and alcohol soluble carbohydrates (70%). It is rich in sugars but devoid of acidity. The tender fruits of lasora are mostly used as vegetable. They are also dried for consumption as vegetable during off-season. Mucilage obtained from half ripe fruits can be used as an alternative to paper glue. Fruits are mostly used for making pickle. Ripe fruits are eaten fresh and used for preparing liquor. Ripe fruits are full of vitamins and regular use supplements hair growth and prevents baldness. Fruits have profound medicinal value and are considered astringent, anthelmintic, diuretic, demulcent and expectorant. It is used in diseases of chest and urinary passage. The lasora kernels are used for curing ring worm. The nutritional composition of lasora is given in Table 12 (Yadav and Goel, 2006).

Table 12 : Nutritional composition of lasora fruit (per 100 g of edible portion)

Constituent	Content	Constituent	Content
Moisture (g)	82.5	Phosphorus (mg)	60
Protein (g)	1.8	Iron (mg)	0.005
Fat (g)	1.0	Potash (%)	1.66
Minerals (g)	2.2	Ash (%)	2.13
Fibre (g)	0.3	TSS (%)	10.2
Carbohydrates (g)	12.2	Acidity (%)	0.2
Calcium (mg)	40	Pectin (%)	4.5

Lasora tree yields high quality wood constituting a good quality timber. It is clear yellow when freshly cut, changing rapidly on exposure to olive or bluish grey and finally to brown or greyish brown in colour and weighs nearly 15 to 18 kg per cubic feet. It is fairly strong, shallowly interlocked-grained, and uneven textured. It seasons well and is liable to insect attack. The wood is said to be more durable in contact with water. The timber saws and works well giving a good surface finish. It is used for boat construction, well-curbs, gun stocks and agricultural implements.

Leaves are used as fodder for goats and cattle during famine and contain about 12 to 16 per cent crude protein and 16 to 27 per cent crude fibre. Leaves are also used for preparation of bowls and plates and as a cheap substitute for wrapping *bidis*. A decoction of the bark is used in dyspepsia and fevers. Lac insect can also be reared on lasora plants. Twigs are used as firewood.

Soil and climatic requirements

Lasora tree is very hardy and can be grown in almost all types of soils. But, well drained deep fertile soil of good water holding capacity is the most suitable (Chandra *et al.*, 1994). It thrives well even on neglected, low fertile, sandy, swampy saline, alkaline soils and hence can be best used for afforestation of the wastelands. It can be grown in sodic soils up to 40 ESP (Exchangeable Sodium Per cent) and in saline soils up to 12 dSm^{-1} (Pathak, 1991). Water logged areas are not suitable for its cultivation. However, lasora may be a good introduction for controlling erosion in degraded areas of high rainfall.

Lasora thrives well under tropical and subtropical climates up to an elevation of 1600 m. It needs warmer climate and is susceptible to frost. But, it can tolerate temperature as low as 1°C for few days and can withstand drought to a great extent tolerating temperature as high as 48 to 50°C during the summer months, particularly in arid regions of Rajasthan. It thrives well in areas receiving an average annual rainfall of 250 to 300 mm.

Cultivars

There are very few named cultivars of lasora. In general, two types of plants *viz.*, large-fruit type and small-fruit types are found growing in different parts of the country. Large fruit cultivars have an average fruit weight of 8.55 g, whereas small-fruit cultivars have fruit weight of 3.0 g. Large fruit types have comparatively more pulp content than small fruit ones. In Gujarat, "Paras Gonda', is a general term for any fruit variety with big sized fruits. Similarly, in Rajasthan 'Puskar Local' is of big size with good fruit shape. At Central Institute for Arid Horticulture (CIAH), Bikaner, Selection-1 has good fruit size, pulp content and productivity. The average annual production of tender fruits is 116 kg per tree. Five genotypes of big fruited lasora were collected and maintained in the field gene bank at Central Institute for Subtropical Horticulture (CISH), Lucknow. The analysis of fruits revealed 11.8 to 16 g weight per fruit, 2.06 to 2.2 g seed weight per fruit, 2.51 to 2.92 cm length and 2.62 to 3.14 cm width of fruit.

Propagation

Lasora is commonly propagated by seed. Seeds are sown in well-prepared nursery beds or in polythene bags (25 x 10-15 cm size) filled with garden soil and FYM mixture in equal proportion. The seed has hard seed coat which does not absorb water easily. Therefore, pre-treatment of seeds with cow dung for 24 hours and soaking in water is advisable to enhance the per cent germination and seedling growth. Watering at 2 to 3 day intervals induces early and uniform germination (Singh *et al.*, 1996). Sowing should be done in June-July. If sown in nursery bed, the depth of seed should be 2 cm in lines spaced about 20 cm apart. Germination starts in about 3 to 4 weeks and is completed in 6 weeks. At lower altitudes, plantable seedlings can be obtained after 3 to 4 months in the nursery, but at higher altitudes, 9 to 12 month old seedlings are ready for planting. Seedlings should be planted in the field during monsoon season. The higher percentage of seed germination can be obtained when seed is treated with H_2S0_4 for 30 minutes. For storage of lasora seeds, ripe fruits are collected from the trees and rubbed to remove the flesh. The healthy stones are dried in the shade and kept in the tin containers. The stones can be stored for about one year in air tight containers. There are 4000 to 7000 stones per kg. Higher percentage of seed germination (91%) was obtained under the shade net house, when the seeds were extracted from fully ripe fruits and quickly sown in the poly bags filled with FYM and soil (1:1) at CIAH, Bikaner (Yadav and Goel, 2006).

Vegetative propagation in lasora has been uncommon till recent past but with the standardization of vegetative propagation technique a new vista has been opened for its commercialization. Shield or T budding can be performed during April-May. T budding, cleft grafting and wedge grafting are also successful (Pundir, 1987; Chovatia and Singh, 1996). T budding was found to be most successful with 96.7 per cent success than other methods. Most suitable period for T budding is the month of August (Chandra *et al.*, 1994). Lasora can also be successfully propagated by patch budding *in-situ* during July-September and by cleft grafting on one year old seedling rootstocks raised in polythene bags during February-March under Lucknow conditions (Shukla and Ali, 2007). Chovatia and Singh (1996) carried out shield budding and wedge grafting at monthly intervals between 10th February and 10th July over 2 years, with a view to comparing different methods. Percentage success was higher with budding (61.07%) than grafting (40.92%) and was highest during April-May. Cleft grafting has been also found successful and best time for grafting is August followed by July and September. Pundir and Porwal (1998) reported 35.5 per cent success with cleft grafting. Trees with small fruits can be converted into big fruit types by top working. Both *Cordia myxa* and *Cordia gharaf* can be used as rootstock for budding and grafting but *C. gharaf* is more suitable due to its hardy nature (Pundir, 1987). In general, the age of the rootstock should be 9 to 12 months, while the scion should be 3-month old.

Planting and after-care

Lasora is planted at 6 to 8 m spacing in square system in pits of 60 × 60 × 60 cm filled with the mixture of soil and FYM in equal quantities. In areas prone to termite infestation, 100 g of chlorpyriphos dust should be mixed with the filling mixture. Digging of pits should be done one month prior to planting during July to September. It is advisable to create small catchments around each plant to harvest run-off water during the rainy season especially in arid and semi-arid areas. One year old seedlings or budded/grafted plants are better for establishment of plants. *In-situ* orchard establishment has also been found successful for rainfed conditions. For the purpose, patch budding during June-July can be done successfully on one year old lasora seedlings.

Systematic information on nutrient requirement of lasora is not available. The amount of fertilizers depends upon age and size of the trees. Each mature tree can be given 25 kg FYM and 500 g nitrogen in the beginning of monsoon. Application of well rotten FYM before flowering and that of nitrogenous, phosphatic and potassic fertilizers after pruning and fruit set in split doses helps in better vegetative growth and fruiting of the trees. In order to achieve vigorous growth and fruiting, application of 10 kg FYM, 100 g nitrogen, 125 g phosphorus and 50 g potassium per tree should be done during monsoon season (Yadav and Goel, 2006). Application of 20 to 30 kg FYM and 1 kg di-ammonium phosphate to five year old trees has been found beneficial in semi-arid areas.

Watering is necessary to young plants after planting at an interval of 15 to 20 days during winter and 8 to 12 days during summer. Monsoon planting requires little watering depending upon the weather conditions. When trees are 2 to 3 year old, watering is not required in general, but conservation of soil moisture in root zone through mulching and other techniques helps the plant in vigorous growth and higher fruit yield. Irrigation should be provided during active growth period of plant, particularly during fruit growth. Hoeing and weeding in the tree basins around the tree trunk and provision of mulching is useful to check the loss of water through soil. Lasoda being deciduous in nature, the dropped leaves can be collected and used as mulch.

Weeds pose serious problem during rainy season especially in the orchards raised in irrigated areas. Weeding should be done as and when required to keep the tree basins weed free. Fully grown up lasora trees need digging around their trunks twice a year *i.e.*, pre-monsoon (June-July) and post-monsoon (October-November).

For erect growth, young plants may be given support. Training is essential to develop strong frame work of the plant. Additional branches should be removed from the main stem and branches be spaced properly. Top should be headed back to increase formation of branches and accordingly 4 to 6 scaffold branches are allowed on the trunk 1 m above the ground level. Lasora does not require regular pruning, however suckers arising from main shoots or rootstocks, undesired, weak and diseased branches should be pruned annually.

Harvesting and yield

Flowering in lasora takes place during March to May in different places coinciding with the emergence of new leaves. The old leaves are shed during winter and trees are leafless for quite some time during early summer. Flowering takes place both in mature and current season's growth in the leaf axils. Lasora tree starts flowering and fruiting 4 to 5 years after planting. Fruits are ready for harvest by the middle of May. Fruits should be harvested at tender stage when fruit is still green and pulp is properly formed for vegetable and pickle purpose. Fruit yield varies with the age of the tree, climate and management practices. In arid regions, lasora tree produces 50 to 100 kg of fruits per tree, while under semi-arid and irrigated conditions an average yield of 100 to 300 kg per tree can easily be obtained.

References

Chandra, A., Chandra, Anju and Gupta, I.C. (1994) Arid Fruit Research, Scientific Publishers, Jodhpur, pp. 235-239.

Chovatia, R.S. and Singh S.P. (1996) *J. Applied Hort.*, **2** (1-2) : 127-134.

Pathak, R.K. (1991) Compendium of Summer School on "Fruit Production and Utilization in Wasteland", NDUAT, Faizabad, pp. 58-64.

Pundir, J.P.S. (1987) Ph.D. Thesis, Sukhadia University, Udaipur.

Pundir, J.P.S. and Porwal, R. (1998) Proc. of Nat. Sem. on Management of Natural Resources of Rajasthan, p. 140.

Shukla, S.K. and Ali, Angrej (2007) *CISH News*, **8**(4), pp. 2-3.

Singh, R.S., Nath, V. and Tewari J.C. (1996) *DECO-MIRROR*, **3** (2): 11-15.

Yadav, P.K. and Goel, Madhu (2006) In: Advances in Arid Horticulture, Vol. II (Eds. P.L.Saroj and O.P. Awasthi), IBDC, Lucknow, pp. 305-318.

Chapter 9 Loquat

Botanical Name	:	*Eriobotrya japonica* Lindl. syn. *Mespilus japonicus* Thunb
Regional/ Vernacular Names	:	Japan plum, Japanese plum, Japanese medlar
Family	:	Rosaceae

Loquat is an evergreen fruit tree grown extensively in the subtropical regions of the world. The fruits are consumed fresh or processed. The seeds are rich in starch.

Origin and distribution

Loquat originated in China (Zhang, 1987). Morton (1987) reported loquat to be indigenous to south-east China. The Dadu River Valley in China is now considered the centre of origin of the genus *Eriobotrya* in China and a large number of communities of *E. japonica.* Since some primitive types of *E. japonica* are found in Japan, some authors consider the origin to be both China and Japan. The loquat has long been grown in Japan and China and several important cultivars have originated in these two countries, particularly in Japan. It is said to have been cultivated in Japan for over 1000 years.

The exact date of its introduction in India is not known. However, Government Botanical Garden (now Horticultural Experiment and Training Centre), Saharanpur, Uttar Pradesh is said to be the first garden to grow it, from where it has spread to different parts of the country (Randhawa and Singh, 1970). The commercial cultivation in India is mostly confined to Uttar Pradesh (Saharanpur, Dehra Dun, Muzaffarnagar, Meerut, Farrukhabad, Kanpur and Bareilly), Delhi, Punjab (Amritsar, Hoshiarpur and Gurdaspur) and Himachal Pradesh (Kangra). It is also grown to a limited extent in Assam, Maharashtra and Tamil Nadu.

Nutritive, medicinal and other uses

The skin of the loquat fruit is easily removed. Peeled and seeded fruits are eaten fresh, sometimes combined with sliced banana, orange sections and grated coconut. They are delicious simply stewed with a little sugar added. The fruits are also used in gelatin desserts or as pie-filling, or are chopped and cooked as a sauce. Loquats can also be canned in syrup. Some people prepare spiced loquats (with cloves, cinnamon, lemon and vinegar) in glass jars. The fruit is also made into jam and, when slightly under-ripe, has enough pectin to make jelly.

The fruit consists of 60 to 70 per cent pulp and 15 to 20 per cent seed. The edible parts of fruit contain 87.4 per cent water, 10.2 per cent carbohydrates, 0.7 per cent protein, 0.3 per cent fat, 0.5 per cent minerals, 0.9 per cent fibre, 0.03 per cent calcium, 0.02 per cent phosphorus and 0.7 mg iron per 100 g edible portion (Rajput and Singh, 1964). The carotenoids of loquat fruits are mainly responsible for flesh and skin colour which varies from yellowish white, yellow to deep orange (Godoy and Amaya, 1995). Total carotenoid values vary widely in fresh fruit peel and pulp and the content in yellow-orange fruit was 5 to 10 times higher than in the yellow-white fruit (Lin and Li, 1985). The fruits are generally used for dessert purposes and also for making jam, jelly, preserves, juice, wine and canning. Chutney is also made from semi-ripe fruits. The loquat jelly is an important product because the fruits are good source of acid and pectin.

The wood is pink, hard, close-grained, medium-heavy and is used for making rulers and other drawing instruments. The young branches are lopped for fodder. The fruit is said to act as a sedative and is eaten to halt vomiting and thirst. The flowers are regarded as having expectorant properties. An infusion of the leaves, or the dried, powdered leaves, may be taken to relieve diarrhoea and depression and to counteract intoxication from consumption of alcoholic beverages. Leaf poultices are applied on swellings. Leaves of loquat are used in curing skin diseases and diabetes. An alcoholic extract has been shown to exhibit anti-inflammatory, antiviral and hypoglycaemic effects (Tommasi *et al.*,1992). Loquats are highly nectariferous, with a heavy fragrance and high honey potential (Yu, 1979).

Soil and climatic requirements

Loquat thrives on wide range of soils, but well-drained, deep, sandy loam soil rich in organic matter is ideal for its cultivation. Oppenheimer

(1947) suggested that instead of very light soils, medium heavy soils with perfect drainage should be preferred. Loquat does not perform well, if the subsoil has gravel, hard pan or excessive lime.

Loquat is highly exacting in its climatic requirements. In India, it grows up to an altitude of 1500 m. A milder subtropical climate with an average rainfall of 60 to 100 cm well-distributed throughout the year is best suited for its normal growth and fruiting. Since flowering takes place from October to late January, at certain places the crop may be destroyed by moderate winter frosts. The plants are permanently injured when the temperature falls to zero and remains for any length of time. The fruit is most susceptible to frost injury when it just starts colouration. In places where summer sets in early and scorching hot winds begin to blow before the fruits ripen, the fruits remain very small and do not mature properly.

Cultivars

The loquat cultivars are grouped into two categories, *i.e.* Chinese and Japanese group on the basis of their origin and time of maturity. In the Chinese group, the fruits are large, pyriform and of deep orange colour, while in the Japanese group, the fruits are small, slender and light coloured. The important cultivars of loquat are Golden Yellow, Matchless, Improved Golden Yellow, Pale Yellow, Large Round, Thames Pride, Fire Ball, Improved Pale Yellow, Mammoth, California Advance, Tanaka, Saharanpur Special, Nectar De Cristal, Parmogi, Centenaria. It is suggested that fruits of an ideal cultivar should be sweet, pulpy, mellow and sub-acidic but pleasant in flavour and possess as few seeds as possible.

The performance of the cultivars namely Golden Yellow, Matchless, Improved Golden Yellow, Thames Pride, Saharanpur Special, Improved Pale Yellow, Pale Yellow, Tanaka and California Advance was assessed at Saharanpur, Uttar Pradesh. Saharanpur Special, Improved Golden Yellow and Improved Pale Yellow proved to be the most suitable for commercial planting because of their earliness and good quality fruits. Tanaka and California Advance ripened late but had good quality fruits (Singh and Lal, 1986).

Propagation

Propagation by seed has been a traditional practice. After extraction from fruits the seeds should immediately be sown as exposure to heat and light results in poor germination. Stratification markedly increased the germination of seeds and embryos of loquat resulting in 98.75% germination (Polat and Kaska, 1992a). Loquat is highly cross-pollinated and it is not advisable to plant seedling trees for commercial cultivation.

Generally, seedlings are raised for use as rootstock. Loquat seedlings are preferred to apple, pear and quince as rootstocks. Quince rootstock causes dwarfing and tolerates heavier and wetter soils than loquat, but in plains of Uttar Pradesh, quince had proved unsuccessful (Singh, 1959). Loquat seeds remain viable for 6 months if stored in partly sealed glass jars under high humidity at room temperature, but the best temperature for storage is 40°F (5°C). They are washed and planted in flats or pots soon after removal from the fruit and the seedlings are transplanted when 15 to 17.5 cm high. When the stem is 1.25 cm thick at the base, the seedlings are ready for grafting.

Inarching has been the common method of vegetative propagation in India. Now-a-days, cleft grafting during January to March is more popular and gives very good success. Loquat can also be propagated by air layering. In this method, 3-month old shoots are ringed and layered. Treatment of ringed shoots with 3 per cent NAA or 2500 ppm IBA (Singh *et al.*, 1961; Srivastava, 1957) has been recommended to improve rooting. Polat and Kaska (1992b) reported 40% as the highest rooting success in air layers with IBA treatment. In China, many farmers successfully propagate loquat by air layering.

Shield budding, using buds from three month old branches in January-February gives encouraging success, while poor success was observed during September. Polat and Kaska (1991) carried out chip, patch and T (shield) budding at 15 day intervals from 15 January to 15 May. March was the most suitable time for budding with 95.48% bud-take. Bud-take was also more than 90% during the period between February and Mid-April. Patch budding was more successful (87.88%) than 'T' (85.43%) and chip (78.10%) budding. The strongest scion shoots were obtained with Akko-13 by chip budding. Polat *et al.* (1999) also found higher percentage bud-take (88.84%) and bud sprouting (63.51%) with scions taken from non-fruiting branches than those taken from fruiting branches.

Planting and after-care

Loquat trees remain productive for 30 years or even more. Thus, location and site selection are important for the planning of orchards. The field should be levelled and pits of 75 x 75 x 75 cm size should be dug well before planting. Generally, a planting distance of 6 × 6 m is followed for loquat. After exposure to sun for 15 to 20 days, the top soil of the pit is mixed with 15 to 20 kg well-rotten farm yard manure and 200 g of single super phosphate. If there is possibility of termite infestation, chlorpyriphos @ 100 g per pit should also be mixed with soil before filling. In these pits, healthy plants are planted either during August or February. For high density planting, 450 to 600 plants per ha can be planted.

Loquat trees produce heavy, luxuriant foliage and large amount of fruits. When the tree begins to bear heavy crops, adequate manure and fertilizers should be incorporated. In India, Singh (1952) suggested that production of 445 kg loquat fruits would require 365 g nitrogen, 360 g phosphorus, 350 g potassium and 325 g of calcium oxide, besides the additional requirement for the vegetative growth. Application of nitrogen at 80 and 120 g per plant to one-year old loquat plants cvs. Golden Yellow and Tanaka in the form of urea proved effective in improving height and spread of plant, number of branches, leaf and leaf area. Calcium deficiency symptoms of loquat were reported by Sato *et al.* (1960), however excess calcium did not show any detrimental effect on tree growth.

Annual fertilizer doses for adult loquat trees in Japan as recommended by Liu (1999) were 30 kg nitrogen, 22.5 kg phosphorus and 24 kg potassium per 1000 m^2; 50% of these amounts (mainly as well rotten organic manure) were applied in autumn and 30% (mainly as chemical fertilizers) in early spring (mid-late February). Sato (1996) suggested standard fertilization with 170 kg nitrogen, 115 kg phosphorus and 125 kg potassium per ha for 10-year old trees of cv. Mogi.

The soil should be kept in good physical condition by thorough cultivation, addition of organic matter and by proper irrigation. Loquat prefers clean cultivation as compared to other methods of orchard management. However, cultivation of leguminous cover crops during winter and summer is advisable (Randhawa and Singh, 1970). The orchard should be kept clean by timely hoeing and weeding. Deep ploughing should always be avoided.

Loquat requires judicious irrigation. There must be sufficient moisture in the soil in order to enable the shoots to develop and the mature terminal buds to fill out properly. As the trees blossom early, the orchard must be irrigated just before the swelling of blossom buds (Randhawa and Singh, 1970). During fruit growth to maturity, maintenance of proper moisture regime by frequent irrigation is advisable. Mulching helps in conservation of soil moisture and enhances fruit yield by 24.4 to 46.42% as compared to control.

Loquat tree is generally trained in central leader or open centre system. It may also be trained into a vase-shape. In the initial stage, stem up to height of 45 to 50 cm should be kept clean. In the fruiting plants, pruning is confined to removal of dead and diseased branches. Harvesting of mature bunch is a kind of pruning and it encourages new growth.

Flowering and fruiting

Loquat bears flowers at the terminal end of current year's growth. The first flush of flowers can be noticed during August-September, however, due to non-viable pollen grains there is no fruit set. The second flush of flowering takes place during October-November and sets maximum fruits. Another flush of flowers can also be noticed during December-January but hot and desiccating winds in early summer have adverse influence and thereby reduce the fruit size. The time of flower bud formation is influenced by the number of leaves on the fruiting shoot (Chang and Lin, 1994).

The problem of gametophytic incompatibility exists in loquat. On the basis of fruitfulness, the loquat cultivars were classified by Singh (1955) into two groups, *i.e.*, self-incompatible cultivars like Golden Yellow, Improved Golden Yellow, Pale Yellow and Large Agra; and partially self-incompatible cultivars which include Large Round, Fire Ball, Thames Pride, California Advance and Tanaka. The cultivar California Advance proved to be the best pollinizer for Tanaka. Tanaka when pollinated with Pale Yellow gave lower fruit set than the selfed ones, indicating partial cross-incompatibility. Golden Yellow, although self-incompatible gave the highest setting when pollinated with California Advance.

Thinning of flowers and/or fruits is an important cultural practice for loquat. Thinning of flowers is necessary to promote the growth of fruits and reduce the size of the inflorescence for bagging. After flower thinning, one inflorescence bears a maximum of ten fruits and later fruits are thinned to reduce the number between 4 and 6 (Fujisaki, 1994). Flower thinning is done by removing as many as 30% of the whole inflorescence and further thinning of flowers from the remaining inflorescence. Another procedure is to remove some lower and upper peduncles and further thinning of flower buds to provide sufficient space for the fruits to develop in good shape and size. Fruit thinning is done as soon as the fruits develop. Fruits of uniform shape and size are retained. The fruits are bagged with paper bags simultaneously with fruit thinning to maintain their attractive appearance.

Harvesting and yield

The fruits should be left on tree until they are fully mature and maximum amount of sugar is accumulated. Loquat attains maturity in about 150 days since opening of flowers. Change of skin colour from green to yellow or orange colour is a useful indicator of harvest stage. Harvesting of immature loquat fruits should be discouraged. All fruits in a cluster usually mature uniformly and, therefore, the entire cluster may be cut. However, in some cases where the fruits at the base ripen before those at terminal end of cluster, care is to be taken to harvest only mature fruits by clipping with stalk. As the fruits are easily damaged, they should be harvested and handled carefully.

The loquat tree starts bearing fruit generally from the third year of planting. The yield increases as the tree grows older and the maximum yield is obtained when the tree attains the age of about 15 years. The average fruit yield is about 16 to 20 kg per tree. However, a healthy well-managed tree produces 30 to 40 kg fruits.

References

Chang, L.R. and Lin, J.H. (1994) *Bull. Taichung District Agric. Improv. Stat.*, **43** : 41-47.

Fujisaki, M. (1994) Chuo Printing Co., Japan, pp. 56-59.

Godoy, H.T. and Amaya, B.R. (1995) *Nutricion,* **45:** 336-339.

Lin, D.Y. and Li, Y.S . (1985) *Acta Horticulture,* **2** : 207 -209.

Liu, C.R. (1999) *South China Fruits,* **28** : 33.

Morton, J.F. (1987) In : Fruits of Warm Climates, (Ed. Julia F. Morton), Creative Resource Systems, Winterville, FL., pp. 103-108.

Oppenheimer, C. (1947) *Agric. Res. Stat. Rehovot Bull.,* No. 44.

Polat, A.A. and Kaska, N. (1991) *Doga, Turk Ormancilik Dergisi* , **15** : 975-986.

Polat, A.A. and Kaska, N. (1992a) *Doga, Turk Ormancilik Dergisi,* **16** : 433-443.

Polat, A.A. and Kaska N. (1992b) *Doga, turk Tarum ve Ormancilik Dergisi,* **16** : 450-459.

Polat, A.A., Durgac, C. and Kamiloglu, O. (1999) *Turkish J. Agric. & For.,* **23** : 1125-1132.

Rajput, C.B.S. and Singh J.P. (1964) *Indian J. Hort.,* **21** : 204-205.

Randhawa, G.S. and Singh, R.K.M. (1970) The Loquat in India, ICAR, New Delhi, p. 62.

Sato, Y. (1996) Cultivation and evaluation of fruit tree, PGR, HCA, Tokyo, pp. 121-129.

Sato, K., Ishihara, M. and Kurihara, A. (1960) *Bull Nat. Inst. Agric. Sci., Hiratseka Ser.*, **8** : *96.*

Singh, M.P. (1952) *Indian J. Hort.,9* **:** 53-58.

Singh, U.P. (1955) *Indian J. Hort.*, **12** : 115-116.

Singh, R.D. and Lal, H. (1986) *Progressive Hort.*, **18** : 314-317.

Singh, Sucha, Singh, K.K. and Chug, D.V. (1961) *Indian J. Hort.*, **18**: 123-129

Singh, J.P. (1959) *U.P. Dept. Agri. Fruit Ser. Bull.* 19, Supdt. Print and Sty., Allahabad.

Srivastava, D. N. (1957) Farm Bull. No. 38, ICAR, New Delhi, pp. 3-20

Tommasi, N., Simone, F. de, Pizza, C., Mahmood, N., Free, P.S., Conti, C., Orsi, N. and Sten, M.L. (1992) *J. Nat. Product,* **55** : 1067-1073.

Yu, D.J. (1979) The Taxonomy of Chinese Fruits (in Chinese), Agri. Press Beijing, pp. 309-316.

Zhang, H.Z. (1987) *World Agri.*, **6** : 54-56.

Chapter 10 Barbados cherry

Botanical Name	:	*Malpighia punicifolia* L. syn. *Malpighia glabra* L.
Regional/ Vernacular Names	:	Barbados cherry, Puerto Rican cherry, West Indian cherry, native cherry, garden cherry, French cherry, acerola
Family	:	Malpighiaceae

The Barbados cherry is a large, bushy shrub or small tree attaining up to 6 m height and an equal breadth with more or less erect or spreading and drooping, minutely hairy branches and a short trunk to 10 cm in diameter. The fruits, borne in the leaf axils, are oblate to round, cherry-like but more or less obviously 3-lobed; 1.25 to 2.5 cm wide; bright-red, with thin, glossy skin and orange-coloured, very juicy, acid to sub-acid pulp. Barbados cherry is well recognized for its vitamin C rich fruits.

Origin and distribution

It is believed to have originated from south America. It is commonly grown in Florida, Puerto Rico and Hawaii. The plant was often planted in school gardens. An explosion of interest occurred as a result of some food analyses being conducted at the School of Medicine, University of Puerto Rico in 1945. The aonla *(Emblica officinalis* L.) was found to be extremely high in ascorbic acid. This inspired one of the laboratory assistants to bring in some Barbados cherries, which the local people were accustomed to eating when they had colds. These fruits were found to contain far more ascorbic acid than aonla, and, because of their attractiveness and superior eating quality, interest quickly switched from aonla to the Barbados cherry. Much publicity ensued, featuring the fruit under the Puerto Rican name of Barbados cherry. Smaller plantings were being developed in Jamaica, Venezuela, Guatemala, Ghana, Philippines and Queensland, Australia, and even in Israel. Many "vitamin C" rich products were prepared and popularized from the Barbados cherry fruits *viz.*, powder, tablets, capsules, juice and syrup.

Nutritive, medicinal and other uses

Barbados cherry is one of the richest natural sources of vitamin C and carotenoids. The cherry-like fruits contain vitamin C up to 4000 mg per 100 g of fresh weight, but typically contain around 1500 mg per 100 g. Generally, green fruits contain twice the amount of vitamin C in mature fruits. The vitamin A content of Barbados cherry fruit ranges from 4300 to 12500 IU per 100 g as compared to approximately 11000 IU for raw carrots. The other vitamins present in fruits include thiamine, riboflavin and niacin. In West Indies and south America, Barbados cherry fruits are commonly eaten fresh or made into jams or jellies. It is an excellent source of antioxidants and are also rich in protein and mineral salts principally, iron, calcium and phosphorus along with considerable anti-fungal activity. The free radical scavenging benefits of Barbados cherry are also useful for preventing abnormal growth and fighting cellular aging. Barbados cherry's mineral salts can aid in the remineralization of tired and stressed skin and the mucilage and proteins can help to hydrate and condition skin. Besides, it can also be used against breast pain and discomfort (mastitis), postmenopausal disorders, diarrhoea, digestive disorders, dysentery; colds, fever, infections, hepatitis and liver pains, stomach ache and prevention of tooth decay. For dessert use, they are delicious merely stewed with whatever amount of sugar is desired to modify the acidity of the particular type available. It is generally used for pulp and juice. The cooked fruits must be strained to remove the seeds and the resulting sauce or puree can be utilized as a topping on cake, pudding, ice cream or sliced bananas. Commercially prepared puree may be dried or frozen for future use. The fresh juice prevents darkening of bananas sliced for fruit cups or salads. It can be used for gelatin desserts, punch or *sherbet*, and has been added as an ascorbic acid supplement to other juices. The juice can be dried and powdered.

The fruits in green stage, medium ripe stage and very ripe stage contain 4500 mg, 3300 mg, 2000 mg vitamin C per 100 g of pulp, respectively. This constituent varies as much as 25% depending upon clone, location, cultural methods and degree of exposure to sunlight during developmental stages and maturity. Treatment of young fruits on the tree with 200 ppm gibberellic acid increases the ascorbic acid content of the mature fruits (Nakasone *et al.*, 1966). The ascorbic acid is not totally destroyed by heat, during jelly preparation and may contain

499 to 1900 mg per 100 g. The bark of the tree contains 20 to 25% tannin and has been utilized in the leather industry. The wood is surprisingly hard and heavy. It refuses to ignite even when treated with flammable fluid unless perfectly dry. The fruits are considered beneficial to patients with liver ailments, diarrhoea and dysentery, as well as those with coughs or colds. The juice may be gargled to relieve sore throat. The composition of Barbados cherry is given in Table 13.

Table 13 : Composition of Barbados cherry (per 100 g of edible portion)

Constituent	Content	Constituent	Content
Moisture (g)	81.9-91.10	Calcium (mg)	8.2-34.6
Protein (g)	0.68-1.8	Phosphorus (mg)	16.2-37.5
Fibre (g)	0.60-1.2	Iron (mg)	0.17-1.11
Fat (g)	0.18-0.1	Carotene (mg)	0.003-0.408
Carbohydrates (g)	6.98-14.0	Vitamin A (IU)	408-1000
Ash (g)	0.77-0.82	Thiamine (mg)	0.024-0.040
		Riboflavin (mg)	0.038-0.079
		Niacin (mg)	0.34-0.526
		Ascorbic Acid (mg)	2000–4500

Soil and climatic requirements

The tree does well on deep and well drained loam, clay loam, limestone and marl soils with slightly acidic pH. The pH should be at least 5.5. Elevation of pH to 6.5 significantly improves root development. Acid soils require the addition of lime to avoid calcium deficiency and increase yield. The lime should be applied to the soil at a depth of 20 cm or more.

Barbados cherry, although classified as a subtropical fruit, thrives well in both tropical and subtropical conditions. The mature trees can survive brief exposure to 28°F (-2.22°C) whereas, young plants are killed by any drop below 30°F (-1.11°C) temperature. It is naturally adapted to both medium and low-rainfall regions, and can tolerate long periods of drought, though it may not fruit until rains start.

Cultivars

In 1956, workers at the University of Florida's Agricultural Research and Education Centre in Homestead, after making preliminary evaluations and selections, chose as superior clone and named the 'Florida Sweet'. The tree was observed to have upright growth habit, large fruits, thick skin, apple-like, semi-sweet flavour, and high yield.

The first promising selections in Puerto Rico, on the basis of fruit size, yield and vitamin C content, were identified as A-l and B-17, but these were later found to be inferior to B-15 in ascorbic acid level and productivity. Yields of 10 clones *viz.* A-l, A-2, A-4, A-10, A-21, B-2, B-9, B-15, B-17, and K-7 were compared over a 2-year period and B-15 far exceeded the others in both years in Puerto Rico (Morton, 1987).

Propagation

Seed should be selected from desirable clones not exposed to cross-pollination with inferior types. They should be cleaned, dried, and dusted with a fungicide. It should also be realized that the seeds in an individual fruit develop unevenly and only those that are fully developed, when the fruit is ripe will germinate satisfactorily. Germination rates may vary from 5 to 50%. Seedlings should be transferred from flat beds to containers when attains 5 to 7.5 cm height. When the seedlings are 2 to 4 months old , they can also be planted to the main field.

Barbados cherry can easily be propagated by air layering, side, veneer, cleft and modified crown grafting, shield or chip budding. However, propagation by cuttings is much easier and popular to raise the plants quickly.

Cuttings of branches, 6 to 12.5 mm thick and 20 to 25 cm long, with 2 or 3 leaves treated with plant growth regulator planted in sand or other suitable media under constant or intermittent mist initiate rooting in 60 days (Nelson and Goldweber, 1956). Later on, these cuttings are transplanted to nursery or containers and kept in shade for 6 months or a year for hardening before being finally planted in the field.

Air layering of shoots treated with seradix/rootex/caredex during spring or summer season is also highly successful and gives 60 to 80% success. Grafting is generally practised only when cuttings of a desired clone are scarce or if a nematode-resistant rootstock is available on which to graft a preferred cultivar; or when top-working a tree that bears fruits of low quality.

Planting and after-care

For planting, pits of 1×1×1 m should be prepared, a month prior to planting. The application of 10 kg well rotten farm yard manure (FYM) or compost per pit is sufficient as basal dressing. The spacing required for planting is 4 to 6 m either way in square system of planting. The plants can be planted any time during the year but the best time is rainy season. Soil should have high water holding capacity with good drainage. Isolated plants or orchard of single cultivar have usually set poor crops, hence mixed planting with different clones should be planted for better return.

Fertilizer requirement of the plant is influenced by various factors like age of the plant, soil and climate along with the crop load in bearing trees. Top dressing with fertilizers at the rate of 500 g of ammonium sulphate, 1 kg of super phosphate and 500 g of potash per plant during its life once in June-July and again in January is found to promote vigour and yield of the plant (Shanmugavelu, 1987). A dose of 400 g nitrogen, 200 g phosphorus and 300 g potassium along with 50 kg FYM per tree per year is recommended for fully grown bearing trees of Barbados cherry (Pandey *et al.*, 2006). The fertilizer should be applied in 30 to 40 cm wide ring made at the radial distance of 100 to 200 cm from the trunk. The total quantity of FYM, phosphorus and half of nitrogen and potassium should be applied after rains and remaining half nitrogen and potassium should be applied just after fruit set.

Irrigation once in four days during the early stages of growth up to one or two years of planting and later on once in 7 to 10 days is necessary. An adequate supply of water is beneficial in promoting good growth and maximum yields of large fruits. It is especially important during the blooming and fruit development period. Irrigation is usually needed during the dry spring months and summer.

The use of mulch is desirable, especially in sandy soils where nematodes pose problem. Mulching helps to conserve soil moisture, control weeds and reduce nematode damage. Straw, grass clippings, leaves, can be used as mulch.

Pruning can be practised for shaping the tree and removing the excessive growth. Some clones such as Florida Sweet tend to grow more upright and open. The upright branches can be headed back to encourage more side branches for developing a good canopy. More bushy selections, producing numerous branches and forming thick growth can be thinned to promote heavier yields. It is more appropriate to prune the plants after fruiting.

Harvesting and yield

For home use, as dessert, the fruits are picked when fully ripe. For processing or preserving, they can be harvested when slightly immature, as they turn from yellow to red. As there is continuous fruiting over long periods, the harvesting is carried out regularly on alternate days in order to avoid any damage to the fruit by falling on the ground (Perez Lopez, 1963).

The fruits are usually picked manually in the early morning, and must be handled with care. People who pick Barbados cherries without gloves and long sleeves may suffer skin irritation from contact with the minute stinging hairs on the leaves and petioles. For immediate processing, some growers shake the tree and allow the ripe fruits to fall on the sheets spread on the ground. Harvested fruits should be kept in the shade until transferred from the field, which ought to be done within 3 hours.

There is great variation in productivity. Individual trees may yield 13.5 to 28 kg in Puerto Rico. In Jamaica, maximum yield in the 6th year is about 36 kg per tree. A plot of 300 trees of 'Florida Sweet' produced 2858 to 23270 kg of fruit from March to November, in Florida.

In Puerto Rico, a planting of 200 trees may be expected to produce 1636 to 2455 kg of juice. From the juice there can be extracted at least 54.5 kg of vitamin C expressed as dehydro-ascorbic and ascorbic acid, providing the content is determined to be 2%. In a commercial operation using ion-exchange resins, the yield of ascorbic acid from Barbados cherry juice is expected to be about 88%.

References

Morton, J.F. (1987) In: Fruits of Warm Climates, (Ed. Julia F. Morton), Creative Resource Systems, Winterville, FL., pp. 204-207.

Nakasone, H.Y., Miyashita, R.K. and Yamane, G.M.(1966) *Proc. Amer. Soc. Hort. Sci.,* **89:** 161-166.

Nelson, R.O. and Goldweber, S. (1956) *Proc. Fl. St. Hort. Soc.,* **69:** 285-287.

Pandey, D., Mishra, D., Pandey, G. and Shukla, S.K. (2006) In : Advances in Arid Horticulture, Vol. II (Eds. P. L. Saroj and O.P. Awasthi), IBDC, Lucknow, pp. 39-48.

Perez Lopez, A. (1963) *J. Agr. U. Puerto Rico,* **47** (3): 193-200.

Shanmugavelu, K.G. (1987) Production Technology of Fruit Crops, SBA Publications, Kolkata, pp.408-410.

Chapter 11 Chironji

Botanical Name	:	*Buchanania lanzan* (L.) Spreng. syn. *Buchanania latifolia* Roxb.
Regional/ Vernacular Names	:	Chironji, piyal, charoli, achar, char, piyar, murkali, nurkale, munga pera, charu, priyalan, thapasa-priya, morala, mudaima, morli, morlu chettu, sara, charwari, cudappah almond, almondette
Family	:	Anacardiaceae

Chironji, a high value minor fruit plant is confined to the forests and community lands of central, western, south-eastern and southern India. It has long been considered a minor forest produce. Traditionally, villagers and tribals harvest chironji fruits from the forest. It plays major role in the rural and tribal economy in parts of Uttar Pradesh, Madhya Pradesh, Chhattisgarh, Maharashtra, Gujarat, Andhra Pradesh, Karnataka, Bihar, Jharkhand and Orissa. The fruit needs domestication and promotion as a commercial fruit crop in suitable areas.

Origin and distribution

Chironji is native to India. Genus *Buchanania* includes trees and shrubs which are distributed in tropical Asia, Australia and Pacific lands. The genus is widely distributed in greater parts of sub-tropics which are frost free. Chironji is commonly found in dry deciduous forests up to an elevation of 1200 m in subtropics and up to 900 m in sub-Himalayan tract. It is a common associate of sal (*Shorea robusta*) and teak (*Tectona grandis*) forest and occupies lower canopy. The species is widely distributed in parts of Uttar Pradesh, Madhya Pradesh, Chhattishgarh, Orissa, Jharkhand, Gujarat, Andhra Pradesh, Karnataka and Maharashtra.

The chironji has not been domesticated as yet. The density of chironji plants in different forest ranges has been worked out by several workers. Plant population was relatively higher near watercourse

(Tewari *et al.*, 2001). Prasad and Pandey (1992) reported a density of 4 to 23.66 per ha in teak dominant forest of Sheoni (Madhya Pradesh) and the density of plants was greatly influenced by its vicinity to habitation. Exact figures of chironji production in India are not available. However, Prasad and Bhatnagar (1993) reported that in Madhya Pradesh and Chhattisgarh alone, chironji seeds were collected to the tune of 1108 tonnes per year.

Nutritive, medicinal and other uses

Chironji is highly valued for its nutritious kernels. Kernels have a pleasant, sub-acidic flavour and are eaten raw or roasted. They form a common substitute for almonds in flavouring sweetmeat, confectionery and betelnut powder. The kernels are often used by tribals in Gujarat as brain tonic. An ointment made from the kernels is used to cure itch of the skin and to remove blemishes from the face. The chemical composition of kernels is given in Table 14.

Table 14 : Composition of chironji kernels (per 100 g of edible portion)

Constituent	Content	Constituent	Content
Moisture (g)	3.0	Calcium (mg)	279.0
Protein (g)	19.0	Phosphorus (mg)	528.0
Fat (g)	59.1	Iron (mg)	8.5
Fibre (g)	3.8	Oxalic acid (mg)	2.0
Carbohydrates (g)	12.1	Magnesium (mg)	373.0
Minerals (g)	3.0	Sodium (mg)	10.2
		Potassium (mg)	436.0
		Copper (mg)	0.86
		Sulphur (mg)	186.0
		Chlorine (mg)	25.0
		Thiamine (mg)	0.69
		Riboflavin (mg)	0.53
		Niacin (mg)	1.5
		Vitamin C (mg)	5.0

It is a rich source of fat and contains 35 to 47.2 per cent light yellow coloured sweet oil with a mild pleasant aroma. Chironji oil is a substitute for almond and olive oil in indigenous medicine. The oil contains 0.14% myristic, 28.9% palmitic, 8.1% stearic, 57.4% oleic and 5.5% linoleic acids. The oil on direct inter-esterification yields a product which may be suitable as a coating material for delayed action tablets. The oil also appears to be a commercial source of palmitic and oleic acids. It is applied to glandular swellings of the neck. Chironji paste is an excellent skin conditioner. Besides fruit, its bark furnishes a natural varnish and is used in tanning in Kerala. Gum exudate from tree trunk is suitable for dressing textile. It is also used in curing diarrhoea, intercostal and rheumatic pains. The leaves are valued for their cardiotonic properties. Leaf powder is a common cure for wounds. Leaves constitute a good feed for cattle. The tree serves as host of rearing Kusumi strain of lac. The settlement of larvae on the twigs is reported to be uniform.

Soil and climatic requirements

Chironji grows well on clayey and laterite soils, but it does not withstand water logging. It prefers red and loamy soils with good soil depth, which are neutral in reaction. In its natural habitat, it withstands absolute maximum shade temperature ranging between 40^0C and 60^0C and the absolute minimum temperature ranging from -1°C to 13^0C and with annual precipitation of 750 mm to 2150 mm. Chironji prefers dry sub-humid climate and is susceptible to frost. Hence, frost prone areas should be avoided for its cultivation.

Cultivars

There are no released cultivars of chironji. The work on conservation and improvement of this fruit has been initiated during last few years. Munde *et al.* (2004) studied variability in 75 genotypes of chironji with respect to characters such as fruit size, fruit weight, mesocarp weight, seed weight, kernel weight, seed coat weight, fruit volume, number of fruits per panicle, number of panicles per tree, number of fruits per tree, total soluble solids, vitamin C and fruit yield per tree and opined that there was a wide scope for selection of superior genotypes for use as improved variety or for use in subsequent hybridization programme. Machewad *et al.* (2003) recorded variability from 62 locations in Parbhani and Nanded districts of Maharashtra and found that NED-1, NED-20, NED-22 and PBN-8 were the most promising genotypes on the basis of various physico-chemical traits. The genetic diversity of 30 chironji genotypes was studied by Singh *et al.* (2006) with respect to various horticultural traits. Out of thirty accessions, CPT-1, CPT-2, CPT-5, CPT-6, CPT-7, CPT-8, CPT-12 and CPT-30 were found promising. Owing to indiscriminate cutting of tree branches by local villagers/ tribals for fruit havesting and felling of chironji trees, the population of trees in its natural habitat is fast depleting. Since it is free for collection by local dwellers, harvesting is often done prematurely with the result that the mature fruits/seeds are not available for its natural dispersal. There is a need to conserve chironji germplasm both *in-situ* and *ex-situ*.

Propagation

Chironji regenerates naturally by seed. Seedlings are also used for rootstocks for vegetative propagation. Seeds should be extracted from fully ripe fruits during April-May. Sowing should be done on raised beds at 2 to 3 cm depth. Seedlings are transplanted to 200 gauge thick perforated polythene bags of 25 x 15 cm filled with soil:FYM mixture in 2:1 ratio at 3 to 5 leaf stage (Singh *et al.*, 2007). Seedlings are ready for transplanting after one year.

Seeds exposed to hot sun quickly lose viability leading to low germination. Shukla and Solanki (2000) reported that 48-hour seed soaking in ordinary water gave as high as 71 per cent seed germination. Mechanical cracking of stony endocarp resulted in 83 per cent germination. However, mechanical breaking is time consuming and poses high risk of damage to embryo. Chironji seed, being recalcitrant, cannot be stored for longer period. Seeds without extraction from the fruits can be stored for 3 months but they lose viability quickly after 3 months. Choubey *et al.* (1997) reported best germination with 0.1 per cent $HgCl_2$ treatment. Bajpai *et al.* (2006) studied the effect of time and depth of sowing on germination and reported that seed sown in March took minimum period to initiate and complete germination at 1 cm depth, while sowing in November registered the maximum period to initiate and complete germination at similar depth. There was no germination in December and January. Similarly, seeds sown at 4 cm depth did not germinate at all. Chironji is a slow growing plant (Ram Prakash, 1998). Tewari *et al.* (2001) reported that chironji is initially a slow grower but its root grows very fast.

Chironji is difficult to root when propagated by cutting. In a study, Singh *et al.* (2002) reported 67 per cent rooting in root cuttings of chironji with 1600 ppm IAA treatment. Best rooting was noticed with 1.5 to 3.5 mm thick roots. Shukla *et al.* (1999) reported that air layering in chironji with the use of IBA up to 5000 ppm failed to initiate rooting. Tewari and Bajpai (2001) reported that patch budding in chironji failed to sprout even after bud take, while chip budding in the month of August showed encouraging results. Tewari *et al.* (2002) reported 80 per cent success through veneer grafting in the months of August-September. *In-situ* grafting is advocated for good success through veneer grafting.

Six month old, slightly mature scion shoots should be selected and defoliated 7 days prior to grafting. One to two year old seedlings are used as rootstocks. Singh (2007) reported from Gujarat that soft wood grafting *in-situ* on 12 month old chironji seedlings performed on 15th July gave highest success and maximum number of leaves.

Planting and after-care

Pits of 1 x 1 x 1 m size should be prepared 8 to 10 m apart in square system of planting during May-June and left exposed to sun for solarization. The planting distance may be reduced to 6 m in poor soils. The pits are filled up with a mixture of top soil and 20 to 30 kg well rotten farm yard manure (FYM). In order to ward off the infestation of termites, pits may be drenched with chlorpyriphos @ 3 ml per litre water. Planting is done with the onset of monsoon after receiving good rainfall during July-August. The seedlings or the vegetatively propagated plants are planted in the centre of pits. Soil around the plant is firmly pressed and irrigation is done immediately after planting. Tewari *et al.* (2001) observed that root length was nearly three times the shoot length, in one year old sapling, leading to poor establishment of plant on transplanting. Shukla *et al.* (1999) observed cent per cent mortality of chironji plants when planted in natural rangelands. However, Shukla and Solanki (2000) reported 93 to 96 per cent survival of plants in field when planted in 60 x 60 x 60 cm pits in shallow gravelly red soil filled with 10 kg FYM coupled with mulching during winter and summer seasons. Frost, rodents and cattle also damage the plants. Orchard establishment *in-situ* is advisable. A survey of plants of varying girths at Madavara forest range, Lalitpur (Uttar Pradesh) indicated maximum canopy spread of 8 m (Tewari *et al.*, 2000).

As regards nutrient management of chironji trees, scientific information is not available. But, a dose of 5 kg well rotten FYM, 100 g nitrogen, 50 g phosphorus and 50 g potassium per plant per year should be applied. This dose should be increased every year up to 10 years of age and a dose of 50 kg FYM, 1 kg nitrogen, 500 g phosphorus and 500 g potassium should be applied every year thereafter. Full dose of manures and fertilizers should be applied in the month of July-August. Pruning in chironji trees is restricted to removal of dried and dead branches only. It does not require regular training and pruning.

Chironji is susceptible to water logging, but natural occurrence of trees in higher densities along the water streams indicates that it requires fair amount of soil moisture for establishment and proper growth.

Irrigations at 10 to 15 day intervals during summer and at 20 to 25 days during winter are recommended. Mulching in basin during post monsoon season helps in moisture conservation and arrests weed growth. Water harvesting measures should be adopted to conserve run off water. Tree basins are kept clean by regular weeding.

Harvesting and yield

Chironji fruits are ready for harvest by April-May. Maturity is indicated by change of fruit colour from green to purple. As soon as fruits start changing colour, the fruits are harvested as the kernels are fully developed at this stage. Harvesting is done manually by shaking the branches. Dropped fruits are collected. Entire bunch of fruits is harvested with a sickle attached to a long bamboo stick/aluminum pipe. Unfortunately, tribals cut the fruiting branches to collect maximum fruits in the shortest span of time. This has led to widespread destruction of chironji trees in forests. The species is not a good coppicer. The existing practice should be discouraged. A fully grown up chironji plant may yield as high as 1.0 quintal of fresh fruits, which in turn will yield 40 kg of stones and 7 to 8 kg kernels. However, 3 to 4 kg kernels per tree are obtained on an average.

After harvest, fruits are dried in sun by spreading in a single layer. If not properly dried, it attracts fungi. Dry fruits are stored in gunny bags in the shade. Storage of fruits in cool moist places is avoided. Thin fleshy pulp with sweet sour taste is edible. For storing large quantities, the pulp is removed. It is advisable to remove the outer covering of pulp by soaking in water and rubbing fruits by hand after it becomes soft. Hard stony endocarp may be dried in partial shade for removing water drops before storage in cool dry place. Long exposure to sun adversely affects seed germination. Stones can be stored safely up to 2 years, while kernels should not be stored for longer periods as they develop bitterness during storage.

References

Bajpai, C.K. Yadav, M.P. and Tewari, R.K. (2006) *Indian Journal of Agroforestry*, **8**(1): 42-50.

Choubey, Archana, Prasad, Ram, Choubey, O.P. and Pant, N.C. (1997) *J. Trop. Forestry*, **13**(11):6573.

Machewad, P.M., Shinde, G.S., Munde, V.M., Kulkarni, N.H. and Prabu, T. (2003) *Orissa J. Hort.*, **31**(2): 110-111.

Munde, V.M., Shinde, G.S., Sajindranath, A.K., Prabhu, T. and Machewad, P.M. (2004) *Journal of Soils and Crops*, **14**(1): 205-206.

Prasad, R. and Bhatnagar, P.(1993) *J. Trop. Forestry*, **9**(3): 188-195.

Prasad, R. and Pandey, R.K.(1992) *J. Trop. Forestry*, **8**(1):62-82.

Ram Prakash (1998) Plantation and Nursery Technique of Fruit Trees, International Book Distributors, Dehra Dun, p. 245.

Shukla, S.K. and Solanki, K.R. (2000) *J. Trop. Forestry*, **16**:44-49.

Shukla, S.K., Solanki, K.R., Kumar, R.V. and Dwivedi, R.P. (1999) *Annual Report*, NRCAF, Jhansi, p.47.

Singh, Jagdish, Banerjee, S.K. and Francis, Alfred (2002) *Indian Forester*, **128** (6): 700-704.

Singh, S., Singh, A.K. and Bagle, B.G. (2007) *Indian Hort.*, **52**(3): 35-36.

Singh, S., Singh, A.K., Apparao,V.V. (2006) *Indian J. Agric. Sci.*, **76**(11): 695-698.

Tewari, R.K and Bajpai, C.K. (2001) *Agroforestry Newsletter*; **13**(1&2):1-2.

Tewari, R.K., Bajpai, C.K. and Chauhan, S.P.S. (2000) *Agroforestry Newsletter*, **12** : 3(1):10.

Tewari, R.K., Kumar, R.V. and Dwivedi, R.P. (2002) *Annual Report*, NRCAF, Jhansi, p. 16.

Tewari, R.K., Shukla, S.K., Solanki, K.R., Bajpai, C.K. and Chauhan, S.P.S.(2001) *Range Mgmt & Agroforestry*, **22**(2) : 255-257.

Chapter 12 Jamun

Botanical Name	:	*Syzygium cuminii* (L.) Skeels syn. *Eugenia jambolana* Lam., *Eugenia cuminii* Druce
Regional/ Vernacular Names	:	Jaman, jambolan, Indian blackberry, black plum, Java plum, jam, kalajam, jambu, jamli, jambul, neereedu, neredum, naval, sambal, nerale, perinnaral, jamo, dinkikudadaru, jammu, phoberkung, thabye
Family	:	Myrtaceae

Jamun is one of the most popular fruits of India grown throughout the country except high hills. It has been popular as an avenue or windbreak tree, planted along roads, canals or around the fruit orchards. During last few years, the fruit has gained commercial importance in both domestic and world markets. As a result, regular orcharding has gained popularity in different parts of country, especially Gujarat and Maharashtra. It is a hardy fruit crop having high medicinal value.

Origin and distribution

Jamun is native to India or East Indies (Singh, 1969). It is widely distributed in tropical and subtropical parts of India, Sri Lanka, Malaysia, Thailand. It is also found in Australia and Philippines. Jamun has successfully been introduced to many other tropical countries like West Indies, east and west Africa, Florida, California, Algeria and Israel.

In India, the maximum number of jamun trees are found scattered throughout the tropical and subtropical regions. It also occurs in the lower ranges of the Himalayas up to an elevation of 1300 m and in the Kumaon hills up to 1600 m. It is widely grown in the larger parts of India from the Indo-Gangetic plains in the north to Tamil Nadu in the south (Chaturvedi, 1956; Singh *et al.*, 1967). It is one of the most hardy fruit crops and can easily be grown in neglected and marshy areas where other fruit plants cannot be grown successfully.

Nutritive, medicinal and other uses

Jamun is a good source of minerals, vitamins, sugars, protein and others. Fruit contains 50 to 65% juice, 9 to 11.5% TSS, 2 to 2.3% acidity, 2.3 to 3.7% pectin, 30 to 40 mg vitamin C and 74 to 100 IU vitamin A per 100 g pulp (Singh, 2001). The nutritive value of this fruit is given in the Table 15 (Singh *et al.*, 1967; Singh and Srivastava, 2000).

Table 15 : Composition of jamun fruits

Constituent	Content	Constituent	Content
Moisture (%)	84.5-86.4	Pectin (%)	2.3-3.7
Protein (%)	0.53-0.65	Vitamin A (IU)	73-100
Fat (%)	0.10	Vitamin C (mg/100 g)	30.3-40.7
TSS (^{0}Brix)	9.0-11.5	Calcium (%)	0.02
Total sugar (%)	5.8-6.9	Phosphorus (%)	0.01
Acidity (%)	2.1-2.5	Iron (%)	0.1

The tasty and pleasantly flavoured jamun fruit is mostly used for dessert purpose and very much liked by people. The fruit is usually shaken with salt before eating. The jamun fruit has sub-acidic spicy flavour. Apart from eating fresh, it can also be used for making delicious beverages, jellies, jam, squash, wine, vinegar and pickles (Ochse, 1961). Jamun squash is a very refreshing drink in the summer season. A little quantity of fruit syrup is useful for curing diarrhoea. A mixture of jamun and mango juices in equal quantity is very good for quenching thirst of diabetic patient. Jamun fruit is used for preparation of wine, particularly in Goa. The vinegar prepared from juice extracted from slightly unripe fruit is stomachic, carminative and diuretic, apart from having cooling and digestive properties. The pomace after extraction of juice contains considerable amount of anthocyanin, sugars and tannin and it can be further utilized in beverage industry. Jamun seed can be used as a concentrate for animals because it is rich in protein, carbohydrates and calcium.

In Goa and the Philippines, jamun is an important source of wine, somewhat like Port, and the distilled liquors, brandy and "jambava" have also been made from the fermented fruit. Jamun vinegar, extensively made throughout India, is an attractive, clear purple, with a pleasant aroma and mild flavour.

The jamun tree is valued in apiculture. The flowers have abundant nectar and are visited by bees *(Apis dorsata)* throughout the day, furnishing most of the honey in the western Ghats at an elevation of

1370 m. The honey is of fine quality but ferments in a few months unless treated.

The leaves are used as fodder for livestock and as food for tassar silkworms in India. In Zanzibar and Pemba, the natives use young jamun shoots for cleaning their teeth. Analyses of the leaves give crude protein 9.1%, fat 4.3%, crude fibre 17.0%, ash 6.0%, calcium 1.3% and phosphorus 0.19%. They are rich in tannin and contain the enzymes esterase and galloyl carboxylase. The essential oil distilled from the leaves is used in flavouring soaps and is blended with other materials in making inexpensive perfume. It consists mainly of mono or sesqui-terpene hydrocarbons which are common in essential oils.

Jamun bark yields durable brown dyes of various shades depending on the mordant and the strength of the extract. The bark contains 8 to 19% tannin and is used in tanning leather and preserving fishing nets.

The wood is red, reddish-gray or brownish-gray, with close, straight grain. The very small, oval pores are often connected by waxy belts of loose tissue. The medullary rays are so fine as to be clearly visible only when greatly magnified. Air-dried wood is apt to crack and split. When kiln dried, the heartwood is hard, difficult to work but polishes well. It is durable in water and resistant to borers and termites; tends to warp slightly. In India, it is commonly used for beams and rafters, posts, bridges, boats, oars, masts, troughs, well-lining, agricultural implements, carts, solid cart wheels, railway sleepers and the bottoms of railroad cars. It is sometimes made into furniture but has no special virtues to recommend it for cabinet work. It is a fairly satisfactory fuel.

The jamun has received far more recognition in folk medicine and in the pharmaceutical trade than in any other field. Medicinally, the fruit is stated to be astringent, stomachic, carminative, antiscorbutic and diuretic. Cooked to a thick jam, it is eaten to allay acute diarrhoea. The juice of the ripe fruit, or a decoction of the fruit, or jamun vinegar, may be administered in India in cases of enlargement of the spleen, chronic diarrhoea and urine retention. Water-diluted juice is used as a gargle for sore throat and as a lotion for ringworm of the scalp.

The seeds, marketed in 7 mm lengths, and the bark are much used in tropical medicine and are shipped from India, Malaya and Polynesia, and, to a small extent, from the West Indies, to pharmaceutical supply houses in Europe and England. Extracts of both, but especially the seeds, in liquid or powdered form, are freely given orally, 2 to 3 times a day, to patients with diabetes mellitus or glycosuria. In many cases, the blood sugar level reportedly is quickly reduced and there are no ill effects. However, in some quarters, the hypoglycemic value of jamun extracts is disclaimed. The dried alcoholic extract of jamun seeds, given orally, reduced blood sugar and glycosuria in patients.

The seeds are claimed to contain an alkaloid, jambosine, and a glycoside, jambolin or antimellin, which halts the diastatic conversion of starch into sugar. The seed extract has lowered blood pressure by 34.6% and this action is attributed to the ellagic acid content. This and 34 other polyphenols in the seeds and bark have been isolated. Other reported constituents of the seeds are protein 6.3 to 8.5%, fat 1.18%, crude fibre 16.9%, ash 21.72%, calcium 0.41%, phosphorus 0.17%, starch 41%, dextrin 6.1%, a trace of phytosterol; and 6 to 19% tannin.

The leaves, steeped in alcohol, are prescribed in diabetes. The leaf juice is effective in the treatment of dysentery, either alone or in combination with the juice of mango or aonla leaves. Jamun leaves may be helpful as poultices on skin diseases. They yield 12 to 13% tannin (by dry weight).

The leaves, stems, flower buds, opened blossoms, and bark have some antibiotic activity. A decoction of the bark is taken internally for dyspepsia, dysentery, and diarrhoea and also serves as an enema. The root bark is similarly employed. Bark decoctions are taken in cases of asthma and bronchitis and are gargled or used as mouthwash for the astringent effect on mouth ulcerations, spongy gums, and stomatitis. Ashes of the bark, mixed with water, are spread over local inflammations, or, blended with oil, applied to burns. In modern therapy, tannin is no longer approved on burned tissue because it is absorbed and can cause cancer. Excessive oral intake of tannin-rich plant products can also be dangerous to health.

Soil and climatic requirements

The jamun tree can be grown on wide range of soils. But, it prefers deep loam and well-drained soil for its optimum growth and good fruiting. Its cultivation should be avoided in very heavy or light sandy soils. Singh *et al.* (1993) evaluated several fruit trees for sodicity tolerance. It was found that within 30 days of planting 80% of jamun trees died at pH 10. Ninety days after planting they showed 40% mortality at pH 9.4.

The jamun is successfully grown in both tropical and subtropical climates with a wide range of temperature and rainfall. Jamun, in general, prefers dry atmosphere at the time of flowering and fruit setting. For proper growth, development and ripening of fruits, early rains are considered beneficial. Trees sometimes grow on riverbanks where roots are exposed to standing water all the year round. It will stand periods of flooding. The young plants are susceptible to cold and drought conditions. Jamun is a hardy fruit crop which tolerates both drought conditions for some time and heavy rainfall (Chovatia and Singh, 2000).

Cultivars

There are no named or standard cultivars of jamun under cultivation. However, considerable variation exists in fruit shape (round, oval, oblong, pyriform), fruit base (flat, necked), fruit apex (flat, pointed), skin colour (deep purple, purple-pink, bluish-black, black), flesh colour (purple, purple-pink, white) and other physical traits of fruits (Singh and Srivastava, 2000). The common jamun cultivar grown under north Indian condition is Ra Jamun. It produces big-sized (length 2.5 to 3.5 cm and diameter 1.5 to 2.0 cm) oblong fruits with deep purple or bluish black colour at full ripe stage. The pulp colour of ripe fruit is purple pink and the fruit is juicy and sweet. The stone is small in size. This cultivar ripens in the month of June-July and is very common in both rural and urban markets.

Another late maturing cultivar bears a small sized (length 1.5 to 2.0 cm and diameter 1.0 to 1.5 cm), slightly round fruit, deep purple or blackish in colour at full ripe stage. The colour is purple, less in juice, weight and sweetness of pulp in comparison with that of Ra Jamun. The stone present in this cultivar is comparatively larger in size. Fruits ripen in the month of August.

A selection known as Narendra Jamun-6 has been identified with desirable traits at Faizabad (Pareek and Sharma, 1993). A seedling selection 'Paras' has been reported from Gujarat with sweet fruits. Some promising lines No. 15, 4, 14 and 13 have been identified from Pune and Ahmednagar in Maharashtra. A promising jamun accession KKV/VGL/SC-3 has been released in the name of 'Konkan Bahdoli' which has bold fruits, small seeds, heavy and cluster bearing habit, high pulp to seed ratio, better table and processing qualities and marketability (Salvi *et al.*, 2006). Kumar *et al.* (1993) observed large variability in jamun fruits of eastern Uttar Pradesh with respect to fruit weight (4.5 to 14.0 g), pulp/seed ratio (2.5 to 10.3), total soluble solids (8.7 to 11.5%) and pectin (2.3 to 3.7%) contents. Study of jamun variability in different areas of Uttar Pradesh and Jharkhand revealed RNC-26, RNC-11, V-8, V-6 and V-7 as promising genotypes on the basis of fruit weight, pulp content and seed size (Patel *et al.*, 2005). Two superior genotypes of jamun CISH J-37 and CISH J-35 were selected by Singh *et al.* (2008).

Propagation

The most common method of propagating jamun is by seed. In order to propagate trees of improved and selected types, vegetative methods of propagation have been suggested. As majority of jamun trees are of seedling origin, they show tremendous variation in their morphology and physico-chemical traits. The enormous variation is due to sexual propagation of the tree and cross-pollination (Priya Devi *et al.*, 2002). Jamun seeds sown fresh usually show a high percentage of germination within two or three weeks, but on storage they lose their viability rapidly (Shanmugavelu, 1967). Mazhar *et al.* (2003) recorded recalcitrant behaviour of jamun seeds. Sasthri *et al.* (2001) recorded higher seed germination (98 to 99%) in large sized seeds than small sized seeds (79 to 89%). Shanmugavelu (1970) reported that treatment with GA_3 improved seed germination and seedling growth. Improvement in shoot length was recorded in response to seed treatment of jamun with cow urine, while root length was maximum with GA_3 treatment at 250 ppm (Swamy *et al.*, 1999).

There is occurrence of polyembryony in jamun (Tiwari, 1926) and the use of apomictic seedlings can be exploited. The percentage of polyembryonic seeds varies from 20 to 25 (Singh and Thakur, 1977). In north India, fruit ripens from June to August. After removing pulp, fresh seeds should be sown 3 to 5 cm deep at a distance of 15 cm in rows made 25 cm apart. The seeds start germinating in 2 to 3 weeks and attain a height of 15 to 25 cm before the winter sets in. The seedlings are transplanted in field/nursery/polythene bag in the following spring (February-March) or monsoon. One year old seedlings are ready for use as rootstocks for budding/soft wood/cleft grafting.

Various methods of vegetative propagation like inarching, veneer grafting, air layering, cutting, patch budding and soft wood grafting have been used for jamun. Saha (1970) recorded 31 per cent success in veneer grafting when one-year old seedlings were used as rootstock. The shoots were taken from spring flush and the operation was done in the month of July.

Soft wood grafting is being commonly used for propagating jamun in parts of Gujarat (Chovatia and Singh, 2000), Karnataka (Madalageri *et al.,* 1991) and Uttar Pradesh. Soft wood grafting using cleft/wedge method has given very good success on one year old rootstocks raised in polythene bags or raised in nursery beds and shifted in polythene bags. Jamun was successfully propagated by patch budding in March and soft wood grafting in July-August under Gujarat conditions using 9 to 12 month old seedlings having pencil thickness as rootstocks (Singh and Singh, 2006; Singh *et al.,* 2007).

Budding is generally employed for propagating jamun in Indonesia. It is an advantageous method of propagation over inarching because of the difficulty in finding suitable scion shoots in jamun near the ground. One-year old, 10 to 12 mm thick rootstock budded in July-August showed better success (Singh *et al.,* 1967). The scion wood or bud wood is taken from one to two year old vigorous shoots. In India, shield, patch and forkert methods of budding are generally employed. Singh *et al.* (1979) at Basti, Uttar Pradesh noted 70 per cent success by patch budding in the month of March. Kumar *et al.* (1993) reported that patch budding preformed during the month of March gave maximum success (72%) than the other methods (shield and forkert budding) and timings (May and July). Chovatia and Singh (2000) reported that in jamun, budding proved superior to grafting when done in June.

Singh (1968) from Saharanpur, Uttar Pradesh reported 86 per cent rooting when white polythene film and 1 per cent IBA in lanolin paste were used in air layering of jamun. Saha (1970) found that 60 per cent of the air layers rooted when treated with 500 ppm IBA during spring and not in the rainy season.

Planting and after-care

Jamun plants can be planted at the distance of 10 x 10 m in square system. Pits of the size 1 x 1 x 1 m are dug up during May-June and exposed to sun for 15 to 20 days. These pits are filled with a mixture of top soil and well rotten farm yard manure (FYM) in 2:1 ratio. The best time of planting is July-August or February-March in north India. The plants should be irrigated immediately after planting. During initial years, young plants should be protected against frost and low temperature during winter by thatching with locally available organic materials, keeping eastern side open for light.

Generally, jamun trees are not manured because they can stand a good deal of neglect. However, an annual dose of about 20 kg of FYM during the pre-bearing period and 50 to 80 kg per tree to bearing trees is considered beneficial. On very rich soils, the trees have a tendency to put on more vegetative growth leading to delayed fruiting. Under such conditions, the trees should not be manured and irrigation should also be given sparingly, withholding in September to October and again in February to March. This helps in fruit bud formation, blossoming and fruit setting. While training the plants the framework of branches is allowed to develop above 1 m from the ground level.

Irrigation is not common in jamun cultivation, but it promotes better growth during establishment and the early stages of growth especially during the dry seasons. In the early age, plants require 8 to 10 irrigations in a year while, bearing trees require 4 to 5 irrigations during fruit development and ripening. In dry areas, the use of water harvesting techniques during rainy season should be used as it encourages subsequent growth and fruiting. As soil moisture is one of the limiting factors for successful cultivation of jamun, use of mulch is very beneficial. It reduces the loss of moisture from the soil, enhances the rate of penetration of rain water or irrigation in the soil and arrests the weed growth. Mulching can be done with black polythene or any organic materials. In general, mulching with rice husk or paddy straw reduces the weed population and conserves the moisture in the soil.

Harvesting and yield

In jamun, new vegetative shoots emerge as terminal growth on the previous seasons branchlets. The new shoots in jamun emerge in two distinct flushes, *i.e.* from February to May and August to October. The flush which appears in the month of February produces maximum growth and flowering. Flower bud differentiation has been observed on 5 to 10 months old branches and it started from last week of January and continued for 43 days (Misra and Bajpai, 1971). It was observed that the flower and fruit drop in jamun start just after opening of flowers and continue up to maturity. This drop can be reduced by two sprays of 60 ppm GA_3, one at full bloom and the other 15 days after initial setting of fruit (Misra and Bajpai, 1971).

The seedling jamun plants start bearing after 8 to10 years of planting while grafted ones after 4 to 5 years. The fruit ripens in the month of June -July. The main characteristic of ripe fruit is deep purple or black colour. The jamun fruit should be picked immediately after ripening. The ripe fruits are picked singly by hand and in all cases, care should be taken to avoid any injury to the fruits. For harvesting, the man climbs up the tree with a bag of cotton on his shoulder. When the bag is full, the picker comes down from the tree and empties them in basket or with the help of rope, hangs the bag down to the ground and another person standing under the tree empties it. The fruits of jamun are generally harvested daily and sent to market the same day. The average yield from a full-grown grafted tree is 60 to 70 kg per year.

References

Chaturvedi, M.D. (1956) *Indian Farming*, **5** : 17-19.

Chovatia, R.S. and Singh, S.P. (2000) *Indian J. Hort.*, **57** (3): 255-258.

Kumar, S., Singh, I.S. and Ashraf, S.M. (1993) *Haryana J. Hort. Sci.*, **22** (1) : 17-21.

Madalageri, M.B., Patil, V.S. and Nalwandi, U.G. (1991) *My Forest*, **27**(2) : 176-178.

Mazhar, A., Khan, M.M., Iqbal, M.J. and Fatima, B. (2003) *Pakistan J. Agril. Sciences*, **40**(3/4) : 164-169.

Misra, R.S. and Bajpai, P.N. (1971) *Indian J. Hort.*, **28** : 273-277.

Ochse, J.J., Soule, M.J. Jr., Dijkman, M.J. and Wehlberg, C. (1961) Tropical and Subtropical, Agricultue, Mac Millan, New York.

Pareek, O.P. and Sharma, Suneel (1993) *Indian Horticulture*, **38**(1): 47-57.

Patel, V.B., Pandey, S.N., Singh, S.K. and Das, Bikash (2005) *Indian J. Hort.*, **62** (3) : 244-247.

Priya Devi, S., Thangam, M., Desai, A.R. and Adsule, P.G. (2002) *Indian J. Hort.*, **59** (2): 153-156.

Saha, A.K. (1970) *Allahabad Fmr.*, **43** : 287-189.

Salvi, B.R., Malshe, K.V., Bendale, V.W., Kshirsagar, P.J. and Jalgaonkar, V.N. (2006) Book of Abstracts, Nat. Symp. on Under-utilized Horticultural Crops held at IIHR, Bangalore on June 8-9, 2006, p.31.

Sasthri, G., Srimathi, P., Malarkodi, K. (2001) *Madras Agric. J.*, **88**(7/9): 524-526.

Shanmugavelu, K.G. (1967) *South Indian Hort.*, **15**: 70-1.

Shanmugavelu, K.G. (1970) *Madras Agric. J.*, **57**: 311-13.

Singh, A.K., Bajpai, A. and Shukla, S.K. (2008) Souvenir, Nat. Sem. on "Sustainable Horticulural Research in India : Perspective, Priorities and Preparedness" held at BBAU, Lucknow on April, 14-15 2008, p. 24.

Singh, R. K. and Thakur, S. (1977) *Proc. Bihar Acad. Agril Sci.*, **25** : 139-142

Singh, A.R. (1968) *Farm J.*, **9** : 27-28.

Singh, G., Bist, J.K., Dagar, J.C. and Singh, N.T. (1993) In : Golden Jubilee Symp. Horticultural Research-Changing Scenario, HSI, May 24-28, 1993, Abstract No.7.5, p. 155.

Singh, I.S. and Srivastava, A.K. (2000) *Indian Hort.*, **45** (3).

Singh, I.S. (2001) *Indian J. Hort.*, **58** (1&2) : 178-182.

Singh, R. (1969) Fruits, National Book Trust, New Delhi.

Singh, R.K., Pandey, I.C., and Prasad, R.S. (1979) *Punjab. Hort. J.*, **19**:74-75.

Singh, S., and Singh, A.K. (2006) *Indian J. Agricultural Sciences*, **76**(4): 242-245

Singh, S., Singh, A.K. and Bagle, B.G. (2007) *Indian Hort.*, **52**(6) : 31&33.

Singh, S.K., Krishnamurthy, S. and Katyal, S.L. (1967) Fruit Culture in India, ICAR, New Delhi, pp. 255-259.

Swamy, G.S.K., Patil, P.B., Athani, S.I., Prabhushankar, D.S. (1999) Advances in Agricultural Research in India, **11:** 89-91.

Tiwari, N.K. (1926) *J. Indian Bot. Soc.*, **5**:124-36.

Chapter 13 Phalsa

Botanical Name	:	*Grewia subinaequalis DC.* syn. *Grewia asiatica Mast.; G. hainesiana* Hole
Regional/ Vernacular Names	:	Phalsa, dhamin, parusha, shukri, phalsi, jana, nallajana, phutiki, palisa, tadachi, buttiyudippe, tadasala, pharasakoli, falsa
Family	:	Tiliaceae

Phalsa is a medium to large shrub or small tree growing to a height of 4.5 m or more, which has long, slender, drooping branches, the young branchlets densely coated with hairs. The alternate, deciduous, widely spaced leaves are broadly heart-shaped or ovate, pointed at the apex, oblique at the base, up to 20 cm long and 16.25 cm wide, and coarsely toothed, with a light, whitish bloom on the underside. Small, orange-yellow flowers are borne in dense cymes in the leaf axils. The round fruits, on 2.5 cm peduncles are produced in great numbers in open, branched clusters. Largest fruits are 1.25 to 1.6 cm wide. The skin turns from green to purplish-red and finally dark-purple or nearly black. It is covered with a thin, whitish bloom and is thin, soft and tender. The soft, fibrous flesh is greenish-white stained with purplish-red near the skin and it gets suffused with this colour in advance stages of ripening. The flavour is pleasantly acid, somewhat like grapes. Large fruits have 2 hemispherical, hard, buff-coloured seeds, 5 mm wide. Small fruits are single-seeded. Phalsa bears delicious fruits. In India, different parts of the plant are used for curing variety of ailments.

Origin and distribution

Phalsa is native to India in Baroda area. There is mention of phalsa in Vedic scriptures. It is commercially cultivated in the states of Punjab, Haryana, Rajasthan, Uttar Pradesh and Madhya Pradesh. Besides these states, it is also cultivated on limited scale in the states of Maharashtra, Gujarat, Andhra Pradesh, Bihar and West Bengal. But, exact acreage is not available. Its cultivation is mostly confined to peri-urban areas around big cities in these states. It is very often grown as filler crop in mango orchards.

Nutritive, medicinal and other uses

The fruits are eaten fresh as dessert, are made into syrup, and extensively used for making soft drinks. The juice ferments so readily that sodium benzoate must be added as a preservative.

The edible portion of phalsa fruit varies from 69 to 93 per cent (Roy *et al.*, 1995). Ripe phalsa fruits are sub-acidic and good source of vitamin A and C. They are fair source of phosphorus and iron. Fruits contain 50 to 60 per cent juice, 10 to 11 per cent sugar and 2.0 to 2.5 per cent acid. The presence of glucose, fructose and maltose has also been reported in ripe fruits. Among the organic acids, citric acid is predominant with traces of malic acid. The vitamin C content ranges between 27.39 and 39.5 mg per 100 g (Singh and Acharya, 1999). The seed coat and the kernel contain 2.0 and 23.7% oil, respectively, which is rich in linoleic acid. The pigment responsible for the colour of phalsa fruit is anthocyanin, *viz.* delphinidin-3-glucoside and cyanidin-3-glucoside. The phalsa seed contains 7.17 per cent crude fat with a typical odour. The composition of fruit is given in Table 16 (Gaurishankar, 2002).

Table 16 : Composition of phalsa fruit (per 100 g of edible portion)

Constituent	Content	Constituent	Content
Moisture (%)	80.8	Vitamin C (mg)	22
Protein (g)	1.3	Ca (mg)	129
Carbohydrates (g)	14.7	P (mg)	39
Fibre (g)	1.2	Fe (mg)	3.1
Fat (g)	0.9	Mg (mg)	72
Total minerals (g)	1.1	K (mg)	351
β-carotene (μg)	419	Na (mg)	4
Niacin (mg)	0.3		

The phalsa fruits are excellent for making juice and squash, however it is mostly used as fresh fruit and has cooling effect. Several beverages, *viz.* nectar, concentrate, squash, and crushed phalsa packed in glass bottles remain acceptable for up to 8 months in cool storage (Roy *et al.*, 1995). The fruits are highly perishable and need to be processed immediately after harvesting (Salunkhe and Desai, 1984, Roy *et al.*, 1995).

The phalsa sticks obtained after pruning can be utilized either for making baskets or supporting vegetable crops. Pruning of one full grown plant is sufficient for making one large or two small baskets. Phalsa baskets are fairly strong and can be used for transporting fruits and vegetables. Ropes can also be made from the fibre obtained from the bark after treatment.

The fresh phalsa leaves are valued as fodder. The bark is used as a soap substitute in Myanmar. A mucilaginous extract of the bark is useful in clarifying sugar. Fibre extracted from the bark is made into rope. The wood is yellow-white, fine-grained, strong and flexible. It is used for archers' bows, spear handles, shingles and poles for carrying loads on the shoulders. Stems that are pruned off serve as garden poles and for basket-making.

The fruit is astringent and stomachic. When unripe, it alleviates inflammation and is administered in respiratory, cardiac and blood disorders, as well as in fever. An infusion of the bark is given as a demulcent, febrifuge and treatment for diarrhoea. The root bark is employed in treating rheumatism. The leaves are applied on skin eruptions and they are known to have antibiotic action.

Soil and climatic requirements

Phalsa can be grown on a wide range of soils, even on moderately alkaline soils. It can be grown with success even in fields where other fruit crops are unsuccessful. However, best results are obtained in well drained, loamy soils. Under waterlogged conditions plants become chlorotic and make poor growth. Phalsa could be grown in wasteland without much care (Nand, 1993) and in un-amended sodic soils having up to 30% exchangeable sodium (Ali and Pathak, 1991). Valia *et al.* (1993) reported the salinity tolerance index for 50% growth depression in phalsa was ECe level of 11.2 dSm^{-1}.

Phalsa is one of the most hardy fruits. It is drought resistant and requires little care. It is a subtropical crop and does best in regions where there is distinct winter and summer. The yield and fruit quality are better in such regions. In regions having no winter, the plant does not shed leaves and produce flowers more than once. The quality and yield are poor in such conditions. Full grown plants can tolerate freezing temperature, for a short period. The plants can tolerate temperature as high as 44°C and high temperature favours ripening of the fruits. It requires clear weather at the time of flowering. Rains at flowering affect setting of fruits.

Cultivars

There is no distinct cultivar available in phalsa. Some growers have, however, given names as 'Local' and 'Sharbati'. Two phalsa types, *i.e.* Dwarf and Tall were recognized, of which the former is more productive (Pareek and Panwar, 1981). The comparative data on growth and fruit characters are given in Table 17.

Table 17 : Characteristics of two types of phalsa (Anon, 1981)

Character	Tall type	Dwarf type
Plant height (meter)	4.5	3.4
Internode length (cm)	101.6	97.6
Leaf size (cm)	20×18	18×15
Lower leaf surface	Light green	Greenish white
Fruit yield (kg)	5.2	3.5
Fruit size (cm)	2.07	2.26
Fruit weight (g)	0.478	0.544
Pulp (%)	81.5	60.3
Juice (%)	5.4	34.6
TSS (^{0}B)	14.0	12.1
Acidity (%)	4.64	3.63

Propagation

Phalsa is commercially propagated by seed. Seedling plants are fairly true to type. Freshly extracted seeds should be used for raising seedlings. Seeds lose their viability under ordinary storage after 90 to 100 days. The viability of seeds can be retained for a period of 6 months under cool storage. Seeds require 15 to 20 days for germination and seedlings get ready for planting in the field by 3 to 4 months. Germination percentage was highest (76%) and the germination period shortest (16 days) with seeds sown immediately after extraction (Singh and Jindal, 1979).

Several vegetative methods are reported to be successful in phalsa. Singh *et al.* (1961) recorded 68.8 per cent success from hard wood cuttings treated with 100 ppm IBA. Singh and Kumar (1967) obtained 72.5 per cent rooting in cuttings when shoots were ringed 15 days before planting and treated with 2500 ppm IBA. A report from Rajasthan Agricultural University, Udaipur claims 76.2 per cent success with hard wood cutting, treated with 250 ppm ethrel for five minutes followed by 3000 ppm IBA for two minutes (Anon, 1981).

The best rooting was obtained following treatment of hard wood cuttings with IBA at 200 ppm (Kathrotia and Singh, 1995). Treatment of hard wood cutting with IBA (2000 and 3000 ppm) significantly increased the root length, number of roots per cutting and percentage survival (Srivastava, 1996). Srivastava *et al.* (1998) observed the best results in terms of growth and survival of phalsa cuttings treated with 2000 ppm IBA when compared with 100 and 300 ppm IBA concentrations. Captan at 0.5% improved survival compared with untreated controls, but sugar solution treatment (5%) had no significant effects.

Ground layering and Air layering have also been tried in phalsa. Randhawa *et al.* (1959) obtained 80 per cent success through ground layering using Seradex-L-30. Mohammed and Chauhan (1970) recorded 85 per cent success, when air layers were treated with a mixture of IBA and NAA at the rate of 1500 ppm each, but when the same mixture was used for ground layering success was 50 per cent only.

Sharma *et al.* (1987) obtained fairly high percentage of success in stooling of phalsa. The plants were headed back 15 cm above the ground level in January. New shoots emerged with the onset of spring. A ring of bark (2.0 cm) was removed from the base of these shoots in August and the shoots were then treated with IBA and NAA in lanolin paste. The shoots were earthed up after 7 days. Rooting occurred in about 45 days. Maximum percentage (79%) of rooting and the longest roots were obtained with IBA 7500 ppm followed by IBA 5000 ppm (72% rooting), while 62 per cent was noted with NAA 10000 ppm.

Planting and after-care

Phalsa can be planted either during July-August or February-March when the plants have shed their leaves. Eight to twelve month old seedlings are better for planting in the field. It also suits as filler crop in mango, guava and aonla orchards and helps in conserving soil moisture with its dense foliage. Usually, planting is done 2.5 to 3.0 m apart both ways, thus, accommodating about 1100 to 1500 plants per ha. However, phalsa is well suited for close planting (high density planting), as it completes flowering and fruiting within 4 to 5 months.

Higher yield per unit area can easily be obtained from phalsa by increasing the plant population. For increasing the plant population, paired row planting system has been tried at Allahabad Agricultural Institute with great success. In paired row planting, two plants are planted at one place at a distance of 0.60 m. The second pair of plants is planted at a distance of 3 m from the first pair of plants. The distance between rows is maintained at 3 m. By adopting this method of planting, the number of plants per unit area is nearly doubled. The total yield is increased by 20 to 30 per cent due to increased plant population.

Phalsa is commonly not fertilized by growers. Moreover, reports on its manuring are also scanty, but it responds to manuring. The fruits are borne on new growth and application of manures and fertilizers encourages vegetative growth. Fifteen kilograms of well rotten farm yard manure applied soon after pruning followed by 125 g nitrogen per plant, just before bud sprouting, was optimum for good yield. Increase in nitrogen level from 0 to 80 kg per ha increased the yield, while further increase of nitrogen dose had negative effect. Application of nitrogen @ 100 g per bush showed significantly higher number of sprouted shoots and fruit yield per ha (Kandolia and Bhuva, 1993). Growth of plants was adversely affected by increasing sodium levels beyond 30.25 ESP (Exchangeable Sodium Percentage) in artificially maintained sodic soils (Ali and Pathak, 1991). Pundir and Pathak (1981) recorded high yield by application of nitrogen (N), phosphorus (P) and potassium (K) at 100, 40 and 25 kg, respectively, per ha. Trials at Rajasthan Agricultural University, Udaipur also revealed that application of 100:40:25 kg of N:P:K per ha gave higher yield (Anon, 1983). Maximum fruit yield in phalsa cv. Sharbati was obtained from plants receiving 100 g N + 40 g P + 40 g K per plant (Singh and Gaur, 1989). As regards tissue analysis studies in phalsa, Bhargava *et al.* (1997) suggested that the fourth leaf from the tip could be considered suitable for reliable nutrient diagnosis of phalsa as it was the youngest mature leaf and fluctuations of macro and micro nutrient contents became stable below fourth leaf. Pareek (1997) also suggested sampling of fourth leaf from tip of shoot 30 days after bud burst. Among the micro nutrients, zinc and iron were found to influence berry size and juiciness. Application of 0.4 per cent $FeSO_4$ alone or in combination with zinc improved the berry size. Application of 0.4 per cent $ZnSO_4$ at pre-bloom stage and after berry set improved the juice content (Singh *et al.*, 1981).

Phalsa is regarded as a drought resistant crop. However, it has been observed that in north India irrigation is essential for obtaining high yield of better grade fruits. Plants do not require any irrigation till January. The first irrigation is needed in the second or third week of February after the application of fertilizer. During March and April, irrigation after every 20 or 25 days seems to be beneficial. However, during May, irrigation after every 15 or 20 days is desirable.

One or two ploughings after pruning of plants is desirable to control weeds and incorporate farm yard manure or compost. If necessary, one ploughing may be given after first irrigation to mix fertilizers and control weeds. The most problematic weeds in phalsa are *Cynodon dactylon, Cyperus rotundus, Euphorbia hirta, Parthenium hysterophorus* and *Mimosa pudica* and Diuron @ 2.0 kg per ha has been recommended for use in phalsa.

Application of growth substances has proved effective in increasing the fruit set and yield in phalsa. Use of cycocel at 250 ppm sprayed twice at an interval of 7 days after 50 per cent fruit set increased the fruit size. It also caused delay in fruit maturity by 3 to 5 days. Application of 1000 ppm ethrel resulted in maximum 71.1 per cent ripening in 5 days after application (Singh *et al.*, 1979). Chundawat and Gupta (1974) recorded improvement in yield and fruit quality with the use of 1000 ppm SADH at early stage of plant growth and it also reduced the plant height. Rema and Sharma (1991) pruned 80 cm high healthy bushes during the third week of January and maintained 7 shoots on each bush. Fruit weight and volume were significantly increased by treatment with 150 ppm NAA or 480 ppm ethrel + 150 ppm NAA during full bloom. A significant increase in fruit total soluble solids was recorded after treatment with 480 ppm ethrel during full bloom, as well as after applying 1920 ppm ethrel or 960 ppm ethrel + 2.5 ppm 2,4-D one week before harvest.

Pruning is one of the important cultural operations in phalsa cultivation and optimum pruning height gives good crop. Annual pruning is essential for securing high yield of better quality fruits as flowers are borne in the leaf axils of new shoots. Plants can be pruned any time during December or January when they are dormant. The phalsa plant is normally pruned at 0.9 to 1.2 m above the ground level in late winter to avoid the possibility of frost damage of the newly emerged shoots (Roy *et al.*, 1995).

Traditionally, phalsa plant is pruned once a year. However, two harvests can be obtained by adopting suitable pruning techniques (Rao and Reddy, 1989; Bhuva *et al.*, 1995). Bhuva *et al.* (1995) observed that double cropping could be possible and profitable by pruning twice, *i.e.* first pruning in the first week of September followed by second pruning during the first week of January. The first crop was harvested in November-December, while the second crop was ready for picking in April-May. In general, the double cropping gave significantly higher yield over single cropping. Treatments with medium and light pruning were at par but gave significantly higher yield over severe pruning, while the fruit weight was noted greater following severe pruning. Light pruning increased the TSS contents in both the cropping system, whereas the acidity was influenced only in summer crop by the severity of pruning.

In north India, pruning should be done once during the second fortnight of January. The bushes should be pruned at 150 cm height from ground level. For rejuvenating old bushes, pruning should be done at ground level. At Bangalore, two crops can be taken in a year, first during April by pruning in September and the other during October by pruning in June (Pareek, 1998).

Harvesting and yield

Flowering in phalsa starts from February-March and continues till May. The flowers are mostly cross pollinated and honey bee seems to play major role in pollination. Fruit weight and size in phalsa cv. Blood Red, increased up to 50 days after fruit set and thereafter declined. Fruits were mature after 60 days, when TSS content was maximum (Gaurishankar, 2002). Phalsa fruits followed a double sigmoid type of growth curve and the respiration curve was of non-climacteric type.

The crop starts ripening by the last week of April. All the fruits do not ripe at the same time. Therefore, picking continues from end of April to first week of June. It is essential to collect the ripe fruits daily from each plant. The ripe fruits are highly perishable and, therefore, they should be marketed as soon as possible.

The average yield from a mature plant in single row system is 6 to 7 kg, whereas the average yield per plant in double row planting is lower (4 to 5 kg) in comparison to single row system. However, the total yield per unit area in double row planting is much higher because of increased plant population.

References

Ali, W. and Pathak, R.K. (1991) *Indian J. Agric. Sci.*, **61**(1) : 54-56.

Anonymous (1981) Ann. Rep. All India Coordinated Fruit Improvement Project, p. 33.

Anonymous (1983) Ann. Rep. All India Coordinated Fruit Improvement Project, pp. 30-31.

Bhargava, B.S., Varalakshmi, L.R. and Kumar, B.P. (1997) *South Indian Hort.*, **45** : 1-2.

Bhuva, H.P., Jadav, R.G., Katrodia, J.S., Patel, R.L. and Raja, K.R.V. (1995) *Indian J. Hort.*, **52** (1) : 19-23.

Chundawat, B.S. and Gupta, O.P. (1974) *Haryana J. Hort. Sci.*, **3** :113-115.

Gaurishankar (2002) In: Fruits Tropical and Subtropical, Vol. II, (Eds. T. K. Bose, S. K. Mitra and D. Sanyal), Naya Udyog, Kolkata, pp. 681-692.

Kandolia, S.H. and Bhuva, H.P. (1993) Abst. Golden Jubilee Symp. on Horticultural Research-Changing Scenario organized by HSI at Bangalore on May 24-28, 1993, Abst. No. 22.22, p. 404.

Kathrotia, R.K. and Singh, S.P. (1995) *Advances in Horticulture and Forestry*, **4** : 35-41.

Mohammed, S. and Chauhan, K.S. (1970) *Indian J. Agric. Sci.*, **40** : 581-586.

Nand, D. (1993) Abst. Golden Jubilee Symp. on Horticultural Research-Changing Scenario organized by HSI at Bangalore on May 24-28, 1993, Abst. No. 7.15, p. 159.

Pareek, O. P. (1997) *Indian Hort.*, **42** (2) : 42-46.

Pareek, O.P. (1998) *Indian Journal of Agricultural Sciences*, **68** (8) : 508-14.

Pareek, O.P. and Panwar, H. S. (1981) *Annals of Arid Zone*, **20** : 281-290.

Pundir, J.P.S. and Pathak, S.P. (1981) Nat. Symp. Trop. And Subtrop. Fruit Crops, Bangalore, p. 59.

Randhawa, G.S., Malik, R.S. and Singh, J.P. (1959) *Indian J. Hort.*, **16** (2) : 119-120.

Rao, L. J. and Reddy, M.G.R. (1989) *South Indian Hort.*, **37** (2) : 115-117.

Rema,J. and Sharma, V.P. (1991) *South Indian Hort.*, **39**(6): 327-331.

Roy, S.K., Khurdiya, D.S. and Waskar, D.P. (1995) In : Handbook of Fruit Science and Technlogy Production, Composition, Storage and Processing (Eds. D.K. Salunkhe and S.S. Kadam), Marcel Dekker, Inc., New York, pp. 553-555.

Salunkhe, D.K. and Desai, B.B. (1984) Postharverst Biotechnology of Fruits, Vol.II, CRC Press, Boca Raton, FL, p. 129.

Sharma, B.B., Saxena, S.K. and Sharma, V.P. (1987) Symp. on Himalayan Horticulture-In the Context of Defence Supplies, DRDO, Tezpur, Assam, October 28-30, 1987, Abst. No. C4, p. 49.

Shrivastava, S.R. (1996) *Recent Horticulture*, **3**(1): 52-54.

Shrivastava, S.R., Sunil Mahajan, Nema, B.K., Amar Chandra, Mahajan, S and Chandra, A.(1998) *JNKVV Research Journal*, 28-29 (1-2) : 80-82.

Singh, D. and Jindal, P.C. (1979) *Haryana J. Hort. Sci.*, **8** : 196-197.

Singh, I.S., Singh, H.K. and Chauhan, K.S. (1979) *Haryana J. Hort. Sci.*, **8** : 196-196.

Singh, J.P., Godara, P.S. and Singh, R.P. (1961) *Indian J. Hort.*, **18** (1) : 46-50.

Singh, R., Bagde, T.R. and Sharma, V.P. (1981) Nat. Symp. Trop. and Subtrop. Fruit Crops, Bangalore, p. 116.
Singh, R.P. and Gaur, G.S. (1989) *Haryana J. Hort. Sci.*, **18** (1-2) : 40-45.
Singh, S.M. and Kumar, H. (1967) *Indian J. Agric. Sci.*, **37** : 151-154.
Singh, S.P. and Acharya, P. (1999) *Horticultural Journal*, **12** (1) : 81-83.
Valia, R.Z., Patil, V.K., Khimani, R.A. and Kapadiya, P.K. (1993) Abst. Golden Jubilee Symp. Horticultural Research-Changing Scenario organized by HSI at Bangalore on May 24-28, 1993, Abst. No. 7.7, p. 156.

Chapter 14 Fig

Botanical Name	:	*Ficus carica* L.
Regional/ Vernacular Names	:	Anjir, anjuru, manjimedi, simayatti, simaiyatti, tenatti, anjura, dumur
Family	:	Moraceae

Fig is one of the oldest fruits cultivated in mild temperate, tropical and subtropical parts of the world. The fleshy fruit is consumed fresh, dried or in processed form. It can also be canned or used for candy or jam making. It is delicious, wholesome and nutritious fruit.

Origin and distribution

The fig is believed to be indigenous to western Asia and distributed by man throughout the Mediterranean area (Morton, 1987). As time went on, the fig-growing territory stretched from Afghanistan to southern Germany and the Canary Islands. It is also thought to be a native of southern parts of the Arabian peninsula, Italy, the Balkan peninsula and the USSR (Tutin, 1964). The common fig is not very common in India in spite of the fact that the soil and climatic conditions prevalent in most parts of the country are most suitable for fig. In India, its commercial cultivation is limited to few states like Maharashtra, Gujarat, Andhra Pradesh and Karnataka. Few scattered trees in home gardens are available in northern states like Uttar Pradesh, Bihar, Punjab and West Bengal. It has bright prospects for commercial expansion in Uttar Prdesh, especially in areas of water scarcity. The total area under fig cultivation in India is about 1000 ha of which 400 ha is in Maharashtra (Singhal, 1998).

Nutritive, medicinal and other uses

Figs are consumed fresh, dried, preserved, candied or canned. Fresh figs are delicious and may be used as dessert or for jam. Great quantity of fig is consumed as dried fruit. Better grades of dried fig are used for making fancy packs, while other grades can be used for the production of alcohol and wine. Figs can be roasted like coffee beans, cooled and packed in containers. Other fig products are spiced or pickled figs, fig bread, fig meat and fig brownies.

Fresh fig is a delicious fruit with high nutritive value. It consists of 84% pulp and 16% skin. Figs owe their food value chiefly to their mineral and sugar contents. The total mineral content is 2 to 4 times that of most other fresh foods. They are richer in iron and copper than nearly all fruits and vegetables and most other dried fruits. Traces of zinc are also reported to be present. Both fresh and dried figs contain appreciable quantities of vitamin A and C and smaller amounts of vitamin B group and D. Total sugars content of fresh figs is 13 to 20% and that of dried of is 42 to 62%. The sugar is present mostly in form of invert sugar. The principal acids in fresh figs are citric and acetic with small amounts of malic, boric and oxalic acids. The acid content ranges from 0.10 to 0.44%.

Dried fig is a high-energy food. Glucose, fructose and sucrose are the major sugars in fruit. The glucose content of the fruit is always more than that of fructose. Iraqi figs contained (per 100 g dry weight) 31.2 g fructose, 34.3 g glucose, 7.82 g sucrose, 1.58 g arabinose and 4.91 g galactose in 79.81 per cent total carbohydrates. The reducing sugar in figs can be increased if the fruit is treated with linolenic acid, which accelerates maturation.

The fruit contains 3.02 per cent (dry-weight basis) total acids (Kucharski, 1964). Citric acid is the predominant acid in fig. Other acids present in small amounts are fumaric, succinic, malonic, oxalic and malic acids. Figs contain most of the recognized vitamins and minerals (Table 18). Drying results in loss of thiamin, riboflavin and niacin. The vitamin

Table 18 : Nutrient composition of fig cultivars (per 100 g of edible portion)

Constituent	Content in cultivars		Mineral	Content in cultivars	
	Calimyrna	Mission		Calimyrna	Mission
Protein (g)	3.00	2.82	Calcium (mg)	174	130
Carbohydrates (g)	58.20	50.10	Iron (mg)	2.5	2.4
Fat (g)	1.9	0.9	Phosphorus (mg)	70	57
Vitamin C (mg)	3.6	3.6	Magnesium (mg)	60	59
Vitamin B1 (mg)	0.079	0.061	Copper (mg)	0.34	0.32
Vitamin B2 (mg)	0.083	0.078	Zinc (mg)	0.48	0.43
Vitamin A (IU)	142	92	Potassium (mg)	682	575
Niacin (mg)	0.71	0.59	Sodium (mg)	10	10

and mineral content varies with cultivar (Bolin and King, 1980). Figs have a distinct flavour that varies considerably depending on the cultivar and/or processing method.

Fruits in both fresh and dried form are valued for laxative properties. It is diuretic, demulcent, emollient and nutritive. It is used in the form of confection and syrups. Medicinal uses such as application against boils and other skin infections have also been attributed to this fruit (Polornin and Huxley, 1965; Font Quer, 1973). Fig helps in maintaining the acid-alkali balance of the body very effectively neutralizing excess acid (Cruess, 1940; Tofu and Tofu, 1969). Fig leaves are used as fodder. Fig latex is used as an anthelmintic. The anthelmintic action has been traced to ficin, a proteolytic enzyme, which has the remarkable power of digesting living helminths.

Soil and climatic requirements

Fig trees grow well on a wide range of soils but do best in deep, non-alkaline clay loams or medium black soils, which are well-drained but retain enough moisture. The fig is one of the most salt and drought tolerant crops (Samson, 1980). It can tolerate a fairly high level of sulphate or chloride salts but it gets injured with the presence of even a small amount of sodium carbonate in the soil. A red loam with a lime substratum and about a meter depth is considered the best for figs in south India (Cheema *et al.*, 1954). Heavy and poorly drained soils usually produce fruits of inferior quality. Soils having high lime content produce fruits of better quality suitable for drying.

The fig is a subtropical fruit. When dormant, mature trees can withstand temperature as low as -12°C to -9.5°C depending on the cultivar but young trees are not very hardy and may need protection during winter. Low temperature below 4.5°C at the time when the trees are not dormant injurious to the plant. Similarly, autumn or spring frosts cause considerable damage to the tree. In north India, where temperature falls to freezing point, the plants remain dormant at that time and normally are not damaged. Figs make good growth when temperature is above 15.5^{0}C to 21^{0}C. Although, there is no need for chilling to break dormancy but buds of most cultivars require some winter chilling.

The climate has great effect on the characteristics of fig fruits also. The size, shape, colour of the skin and pulp, quality and tendency towards parthenocarpy are markedly affected by climate. Best quality figs are produced in regions with dry climate, especially at the time of fruit development and maturation. High humidity coupled with low temperature usually results in fruit splitting and poor fruit quality. Fruits formed during March-April under Pune conditions do not ripen because of high humidity and low temperature. Fruits tend to ripen prematurely at temperatures above 38°C and such fruits have tough skin and low pulp content. Fresh fruits exposed to hot winds in warmer areas, though a little sweeter, remain small in size. On the other hand, mild temperature results in large and succulent fruits.

Cultivars

Depending upon the nature of flowers and the method of pollination, there are four distinct classes of fig, *viz.* Common fig or Adriatic fig, Capri fig, Smyrna fig and San Pedro fig.

Common fig

In this type, the individual flowers are pistillate and the fruits develop without the stimulation of pollination and fertilization. Some cultivars of this type are Mission, Adriatic, Brown Turkey, Celeste and Conadria.

Capri fig

It is the most primitive cultivated type with short-styled pistillate flowers and functional staminate flowers. Most capri figs are not edible but are grown because they harbour a small wasp, *Blastophaga psenes*, which is necessary for pollination and fruit setting.

Smyrna fig

In this type, the fruits do not develop unless the flowers are pollinated with pollen carried from the male flowers of the capri fig by the tiny *Blastophaga* wasp. Calimyrna is the most common and widely grown cultivar of Smyrna type.

San Pedro fig

It is an intermediate type in which the first crop known as 'breba' crop, is completely parthenocarpic and does not require pollination and fertilization of flowers, but the second crop develops only if the flowers are pollinated as in the case of Smyrna fig. San Pedro, King and Gentile are common cultivars of this type.

Of these four types, Smyrna fig is commercially most important and is grown extensively in Europe and the USA. In India, common fig is mostly grown. It is considered to be a hybrid between imported *F. carica* and indigenous species. A large number of cultivated forms are grown in which the fruits vary in shape, size, colour of skin, colour and flavour of flesh and period of ripening. Some of the cultivars grown in India are Poona fig, Black Ischia, Brown Turkey, Turkish White, Kabul and Marseilles. The figs grown in many parts of India are named after locality, but they do not exhibit any special distinction to warrant varietal names. Poona fig is of medium size, bell-shaped and light purple in colour with rosy flesh. In south India, Poona fig and Marseilles are commonly grown, the later thriving well in the hilly regions. The fruits of Marseilles are medium-sized, pale green on the rind with a whitish sweet flesh. Some local selections like Shahi, Maisram, Hindupur and Daultabad are also cultivated in some parts of the country. At Bangalore, Poona fig and the exotic hybrids like Deanna, Conadria, Zardrinz and Excel were evaluated with Brown Turkey as rootstock (Pareek and Vishal Nath, 1996). All these varieties were parthenocarpic and bore fruits having prominent rib, separable skin and white spots on surface. The fruit size was largest in cultivar Deanna and smallest in Excel. The fruits of Deanna and Conadria were sweet whereas those of Poona and Excel are slightly acidic.

Propagation

The fig is easily propagated by hard wood cuttings, although budding, grafting and air layering are also successful. Hard wood cuttings, 20 to 30 cm long and 0.5 to 0.7 cm thick, taken from 2 to 3 years old mature shoots and planted 30 cm apart both ways in prepared nursery beds at the beginning of rains. After a year, these are ready for transplanting. Amonov (1972) reported details of propagation techniques of figs by cuttings. He found the optimum length and diameter of cuttings being 30 to 40 cm and 1.1 to 1.5 cm, respectively. Cuttings taken from the base of the shoot and lower part of the crown rooted better than those taken from the middle part of the shoot or from the upper part of the crown. Rooting was best in cuttings from 3-year old wood. The optimum soil water for rooting was 20 to 22 per cent and for shoot growth 16 to 18 per cent. Application of nitrogen @ 30 kg and and phosphorus @ 22 kg per ha resulted in increased growth of transplants. Pre-girdling of fig canes, 30 days before taking cuttings, resulted in quick and increased rooting (Nalawadi and Sulikeri, 1972). In north India, cuttings are made during January-February at the time of pruning, while in south India cuttings are usually made and planted in the rainy season. Sometimes cuttings are directly planted in the field during June-July rather than growing them in the nursery in the first instance. Planting long cuttings (1 to 1.25 m) directly to the permanent location is also quite common in European countries (Condit, 1947).

Figs can be air layered successfully and monsoon is the best time for it (Nagpal and Puri, 1986). One-year old branches if layered in June can be planted in the field by August-September.

Figs can also be propagated by means of budding or grafting. Propagation by budding and grafting enables use of rootstocks resistant to adverse conditions. Chip budding, patch budding and wedge grafting during October gave the success of 64.8, 53.3 and 45.8 per cent, respectively (Pareek and Vishal Nath, 1996). Success with side grafting was also reported (Naik, 1949). Fig is quite susceptible to root knot nematode and use of resistant rootstocks, such as *Ficus glomerata,* offers solution in areas where nematode is a problem. Top working can be done to change an inferior type cultivar into a better one. For this purpose, shield or patch budding, or cleft or bark grafting can be practiced (Nagpal, 1966). Grafting of figs on such species as *Ficus glomerata, Ficus hispida, Ficus palmata* and *Ficus religiosa* has given success in some parts of the country. At Bangalore, vigorous cultivar Brown Turkey was used as rootstock (Pareek and Vishal Nath, 1996) for the exotic hybrids Deanna, Conadria, Zardrinz and Excel along with Poona fig. The hybrids which otherwise did not grow well, put forth good growth and started bearing on the rootstock.

Planting and after-care

In northern India, rooted cuttings are transplanted in early spring when the plants are still dormant but in western India, monsoon is the best planting season, while in southern India, planting can be done any time during the year, but August-September is most preferred. Planting is done mostly in square system. Generally, spacing of 8 x 8 m is used, but it may be increased to 10 m. Close spacing at about 4 m, both ways, is adopted in Tamil Nadu and Andhra Pradesh (Nagpal, 1966). In some places of south India, the plants are spaced as close as 2 to 3 m apart. It is, however, recommended that cultivars ordinarily grown in India should be planted 5 to 7 m apart to obtain the best yields. In cv. Poona, Keskar (1985) suggested 5 m planting distance both ways and training of three or more stems for higher production.

Pits of 60 × 60 × 60 cm are dug at least a month before planting. Pits are usually filled with a mixture of farm yard manure or compost and garden soil in equal proportion before the onset of monsoon in southern and western India and in the beginning of spring in northern India. Water is applied immediately after planting.

Proper doses of manures and chemical fertilizers should be applied in fig orchard to obtain quality production. One year old trees should be supplemented with 9 kg of farm yard manure (FYM) and 455 g of neem cake or 35 g of nitrogen in the form of urea (Singh *et al.*, 1963). The addition of manures should be increased progressively at the rate of 7 kg FYM, 450 g neem cake or 35 g nitrogen every year. A 5-year old fig tree should receive about 40 kg of FYM, 2 kg of neem cake and 150 g of nitrogen. In fig, nitrogen is essential for vegetative growth and reproductive development, phosphorus for fruit colour and maturation and potassium for fruit yield and quality (Harai, 1964 and Haggag and El-Shawy, 1987). At, Rahuri, application of 900 g nitrogen, 250 g phosphorus and 250 g potasssium to each 6 to 7 year old tree of Poona fig planted at 5 × 5 m spacing was found optimum (Pareek and Vishal Nath, 2006).

Fig is fairly drought resistant and it is rarely irrigated in many fig growing countries. However, it has been reported from Kairo that irrigated plants show greater shoot length and yield of fruits. Fruits from irrigated trees are heavier by 13.5 to 16 per cent (Ezzat *et al.*, 1975). In India also, fig is irrigated only during dry months. Singh *et al.* (1963) suggested irrigation every 10 to 12 days during summer months. However, care should be taken that at the time of fruit ripening, no irrigation is given as it may result in the production of insipid fruits. Excessive water in the soil may also cause splitting of fruits. Drip irrigation system, if employed saves irrigation water significantly, besides improving the fruit quality.

Pruning is necessary to stimulate production of new growth, which bears fruits and training is required to produce strong framework capable of bearing heavy crop and to facilitate cultural operations as well as harvesting of the crop (Nagpal, 1966). Pruning increases yield and fruit weight, and extends ripening period. Early cultivars respond better than late cultivars (Kazas, 1979). In Egypt, moderate pruning and heading back by one half were found to be the best (Khalil *et al.*, 1982). To control excessive growth, terminal pinching was found useful (El-Kassas *et al.*, 1988a). Ringing the shoot at the base between the second and third nodes in mid-June gave added advantage (El-Kassas *et al.*, 1988b).

Harvesting and yield

For best flavour, the fruits should not be picked until they are soft and wilt at the neck, hanging down due to their own weight. They are then fully mature and ripe and ready to be eaten. Milky latex exuding from the stem when the fruit is pulled off indicates that the fruit is still immature.

As fresh figs are extremely perishable, fully ripe fruits cannot be transported to distant markets. For the purpose, it is better to harvest the crop when slightly immature. Ripe fruits are picked either from the tree by twisting the neck at the stem-end or by cutting it or gathered after they drop on the ground. Harvesting may start from the middle of February and continue till June. Fruits are picked every 2 to 4 days. The average yield is 180 to 360 fruits per tree A harvest of 12 tonnes per ha can be obtained from a well managed orchard (Samson, 1980).

References

Amonov, K.H. L. (1972) *Subtrop. Kul'tury*, **6** : 91-107.

Bolin, H. R. and King, A. D. (1980) In: Figs, Tropical and Subtropical Fruits: Composition, Properties and Uses (Eds. S. Nagy and P. E. Shaw), AVI, Westport, CT, pp. 492-505.

Cheema, G. S., Bhat, S. S. and Naik, K. C. (1954) In: Commercial Fruits of India, Macmil1an & Co., Kolkata.

Condit, I. J. (1947) The Fig, Chronica Botanica Co., Waltham, Mass, USA.

Cruess, W. V. (1940) *Fruit Prod.* **19** : 230.

El-Kassas, S. E., El-Sese, A. M., El-Agomy, S. Z. and Mohammed, E. A. (1988a) *Assuit. J. Agric. Sci.*, **19** : 355.

El-Kassas, S. E., El-Sese, A. M., El-Agomy, S. Z. and Mohammed, E. A. (1988b) *Assuit. J. Agri. Sci.*, **19** : 339.

Ezzat, A. H., Naguib, M., Elemenchawy, E. A., Mitwalli, S. and Hassan, G. F. (1975) *Agric. Res. Rev.*, **53** : 15-18.

Front Quer, P. (1973) In: *Plantas Medicinales El Dioscoridges Removadol Editorial Labor*, Barcelona.

Haggag, M. N. and El-Shawy, H. A. (1987) *Alexandria J. Agric. Res.*, **33**: 19.

Harai, T. (1964) *J. Jpn. Soc. Hort. Sci.*, **33** : 273.

Kazas, A. N. (1979) *Subtropicheskie Kil'tur*, **4** : 78.

Keskar, B. G. (1985) Proc. 3rd Nat. Workshop on Arid Zone Fruits. MPAU, pp. 155-156.

Khalil, F. A., El-Said, M.E, EL-Said, V. F., Noaman, V. F. and El-Sherif, A. H. (1982) *Agr. Res. Rev.*, **60** : 23.

Kucharski, S. (1964) *Polska*, **20** : 581.

Morton, J. F. (1987) In: Fruits of Warm Climates, (Ed. Julia F. Morton), Creative Resource System, Winterville, FL., pp. 47–50.

Nagpal, R. and Puri, S. (1986) *Indian. J. Forestry*, **9** : 232-236.

Nagpal, R. L. (1966) Fig Cultivation in India, Farm Bull. No. 42, ICAR, New Delhi.

Naik, K. C. (1949) In: South Indian Fruits and Their Culture, P. Varadachary & Co., Chennai.

Nalawadi, U. G. and Sulikeri, G. S. (1972) *Indian J. Hort.*, **29** : 283-286.

Pareek, O.P. and Vishal Nath (1996) Coordinated Fruit Research in Indian Arid Zone : A Two Decade Profile (1996-1995), pp.75-79.

Polornin, O. and Huxley, A. (1965) Flowers of the Mediterranean, Chatto & Wincus, London.

Samson, J. A. (1980) In: Tropical Fruits, Tropical Agriculture Series, Longman, London and New York.

Singh, S., Krishnamurthy, S. and Katyal, S. L. (1963) Fruit Culture in India, Indian Council of Agricultural Research (ICAR), New Delhi.

Singhal, V. (1998) Handbook of Indian Agriculture, Vikas Publishing House Pvt. Ltd., New Delhi, p.186.

Tofu, M. and Tofu, O. L. (1969) *Japanese Patent*, **69** : 12, 747.

Tutin, T. G. (1964) Flora Europaea, Vol. 2, Cambridge University Press, London.

Chapter 15 Mulberry

Botanical Name	:	*Morus* species *Morus alba* L. (White mulberry) *Morus rubra* L. (Red mulberry) *Morus nigra* L.(Black mulberry)
Regional/ Vernacular Names	:	Tut, tutri, chinni, ambat, shetur, shahtoot, reshmi chattu, pippalipandu chattu, musukette, kambli chedi, Hipnerle, tuto, tuticoli, siah tut, kimbu, bola, dieng-bylliet, rokseng, hmubelbing, ayumasing, kimu, himu, karum, karttut, tuti, dieng-soh-tungkhar
Family	:	Moraceae

Mulberry is one of the most common fruit trees found all over India. It is a perennial, fast growing fruit tree which thrives well under tropical, subtropical and temperate regions of India. It is chiefly cultivated for rearing silkworm, but it has immense potential to be popularized as a commercial fruit crop. Mulberry fruits are rich in antioxidants. It has not been able to attain commercial significance due to its perishable nature. The genus Morus belongs to family Moraceae. There are more than 100 species under the genus Morus, among which, *Morus alba*, *Morus rubra* and *Morus nigra* are the main species. Asia is considered to be the centre of origin of mulberry. There are three major species of mulberry; *Morus alba* L. (white mulberry), *Morus rubra* L. (red mulberry) and *Morus nigra* L. (black mulberry) originated in China, north America and Iran and India, respectively.

Nutritive, medicinal and other uses

Ripe mulberry fruits are known to have 85 to 88% water, 7.8 to 9.2% carbohydrates, 0.4 to 1.5% protein, 0.4 to 0.5% fat, 1.1 to 1.9% free acids, 0.9 to 1.4% fibre, 0.7 to 0.9% minerals. Singh (2001) reported that mulberry fruits have 100 per cent edible portion with 9 to 11% TSS, 0.8

to 1.1% acidity, 12 to 20 mg per 100 g of the vitamin C and 0.5 to 0.7% minerals. The ripe fruit of mulberry is highly valued for its delicious taste. The fruits are consumed either fresh or after extraction of juice. The ripe fruits are processed or conserved throughout Asia particularly in Central Asia and in the near east. The dominant taste of the ripe fruit is sweet but usually with sub-acidic blend due to the high water content and low level of other flavouring ingredients. The fruit is very delicious and used for chutney preparation at immature stage and for table purpose on ripening. The flavour improves, if the fruit is dried; it can then be used as a substitute of raisin. The ripe fruits of mulberry contain about 9% sugar with mallic and citric acid . The composition of mulberry fruit is given in Table 19 (Srivastava and Kumar, 1994).

Table 19. Composition of mulberry fruit (per 100 g of edible portion)

Constituent	Content	Constituent	Content
Water (%)	86.5	Phosphorus (mg)	30
Protein (%)	1.1	Iron (mg)	2.3
Fat (%)	0.4	Vitamin C (mg)	12
Minerals (%)	0.6	Carotene (µg)	57
Dietary fibre (%)	1.1	Thiamine (mg)	0.04
Carbohydrates (%)	10.3	Riboflavin (mg)	0.13
Calcium (mg)	70	Niacin (mg)	0.5

The fruits of *M. alba* are eaten fresh or made into juice, stews and tarts. They may be squashed and fermented to yield spirituous liquors. Ripe fruits of *M. nigra* are well flavoured and are eaten fresh or made into jam, jelly and *sharbat*. Singh (2001) reported that mulberry fruits may be used of preparing jam, squash and syrup. In recent years, mulberry fruit juice has been promoted as a health beverage, and it has become very popular in China, Japan, and Korea. Without adding preservatives, the original juice of mulberry fruit remains fresh under cold storage for 3 months, while the bottled beverage remains fresh for 12 months at room temperature. Mulberry fruits are rich in anti-oxidant compounds.

The most important use of white mulberry is use of its leaves for rearing of silk worm. They are also useful as cattle fodder. Being nutritious and palatable, they improve milk yield when fed to dairy animals. Analysis of leaves of *Morus alba* gave 14% protein, ether extract 6.8%, N-free extract 49.7%, total ash 13.8%, calcium 2.74% and phosphorus 0.45% on dry weight basis. Feeding experiments have shown that up to 6 kg of leaves per day can be fed to milch cows without affecting the health of animals adversely.

Mulberry fruits are rich source of antioxidant compounds. Shivashankara *et al.* (2006) recorded higher antioxidant values in dark red and purple fruited mulberry as compared to white type. Dark red fruited mulberry was rich in phenols, anthocyanin, ascorbic acid and flavonoids. Mulberry fruit juice is used for curing throat infection, controlling fever and also for blood purification. Bark is used for making nerve tonic. Mulberry leaves are considered diaphoretic and emollient. A decoction of leaves is used as a gargle in inflammations of the throat. The fruit is cooling and laxative and it is used for sore throat, dyspepsia and melancholia. The fruit juice forms a grateful drink during convalescence after febrile diseases. The root is reported to possess anthelmintic and astringent properties. The bark is used as a purgative and vermifuge.

The mulberry wood is used for making agricultural implements, table, chair, boat and sport items like hockey sticks and tennis rackets. Pruned twigs are used for basket making in the Indian Subcontinent. Bark is used for making good quality paper. The stem bark of *M. alba* is fibrous and has been employed in China and Europe for paper making. The bark may be stripped from waste branches, after separating the leaves for silkworm feeding, and worked into paper pulp. Mulberry trees are utilized for landscaping in Asia, Europe and southern USA (Tipton, 1994). Because of low water requirements and amenability to pruning mulberry trees are suitable for urban conditions, house gardens, street shade and city embellishment.

Soil and climatic requirements

The soil for mulberry cultivation should be deep, well drained, clayey loam in texture, friable structure with good water holding capacity. Slightly acidic soils, free from injurious salts are ideal for mulberry cultivation. Saline and alkali soil though not suitable for mulberry cultivation but can be utilized up to 30 to 40 ESP and 9 to 12 dSm^{-1}.

Mulberry can be grown from sea level to an elevation of 1000 m above mean sea level. In northern parts and temperate regions of the country, mulberry tree undergoes dormancy during winter, while in the tropics the growth is continuous. In Karnataka, where temperature ranges between 21^0C and 31^0C and West Bengal with the temperature ranging from 15^0C to 37^0C, mulberry exhibits good growth throughout the year due to favourable temperatures. But under extremely low (below 13^0C) or high (above 38^0C) temperatures the sprouting of buds and growth of mulberry are affected. In areas of 600 mm or less rainfall, the growth is limited due to moisture stress resulting in low yields. The atmospheric humidity in the range of 65 to 80% is ideally suited for mulberry growth.

Cultivars

White mulberry is commonly grown for rearing silkworm and rarely for its fruits. It is rather uncommon to grow mulberry for fruits and it has been neglected as a fruit crop, in spite of immense potential. During last few years, work has been started in some institutions on collection and conservation mulberry genotypes for popularizing it as a fruit crop. Promising mulberry genotypes have been collected at Indian Institute of Horticultural Research (IIHR), Bangalore and Central Institute for Subtropical Horticulture (CISH), Lucknow. Ten genotypes of mulberry collected at CISH, Lucknow are under evaluation. Out of these, two genotypes, *viz.* CISH Mulberry-2 and CISH Mulberry-8 have been found as the most promising in terms of fruit size, TSS, colour and appearance (Shukla, 2008). Black or dark red fruited mulberry types with high total soluble solids (TSS) and large fruit size are ideal for promoting its cultivation as a fruit tree. Planned mulberry orchards exists in the state of Uttar Pradesh. As the cultivation of this crop is highly remunerative, it is gaining popularity day by day.

Of the several cultivars of black mulberry developed in Europe only Black Persian has been the most popular which has juicy, black plump fruits of 4 × 2.5 cm size with a rich sub-acidic flavour. The varieties grown for fruit in the USA are Trowbridge Thorburn, Ramsey White, Victoria Munson and Merit, which produce large fruits with good flavour (Ottman, 1987).

Propagation

Mulberry can be propagated by seeds or vegetatively through stem cuttings. The stem cutting is the most common method of propagation in India because of the distinct advantages like speedy multiplication of parent material and maintenance of the desired characters of the plant. In temperate climates like Japan, South Korea and Jammu and Kashmir state in India, the improved varieties of mulberry do not root readily, therefore, root grafting is popular for commercial cultivation.

Seeds of mulberry are orthodox and can be stored under dry cool conditions. Soaking of seeds in 100 ppm GA_3 for 24 hrs results in higher seed germination (Rao *et al.*, 1963). Seeds are sown during June in well prepared nursery beds. After attaining 15 to 20 cm height, seedlings are shifted to earthen pots or polythene bags. For seed germination, certain prerequisites such as selection of quality seed, preparation of land and seeds are needed. The fresh seeds will have higher germination rate than the stored ones. Seeds should not contain flesh and can be obtained by repeated washing with fresh water. After extraction the seed should be well dried. A minimum level of six per cent moisture in seed should be maintained for optimum germination on sowing. Seeds should be sown by broadcasting or in line in well prepared beds. After germination the seedlings are shifted to polythene bags, pre-filled with a mixture of orchard soil, compost and sand in equal proportions. During rainy season, plant should be transplanted in well prepared field. Seed propagation is mainly for raising hybrid populations.

Mulberry is by and large propagated by semi-hard wood cuttings. The 6 to 8-month old mature shoots with 15 to 20 cm length, 10 to 12 mm diameter and 3 to 4 healthy buds are selected for cutting. The cut ends of the stem cuttings should be clean at an angle of 45° with a sharp knife without bark split. Cuttings are potted in sand beds or pots during winter. They start sprouting with increase in temperature during February-March which can be shifted to polythene bags/earthen pots for growth. Irrigation is done at regular intervals. With the onset of monsoon season, saplings are ready for planting. Semi-hard wood cuttings give better results in respect of sprouting and survival.

Mulberry can also be propagated by budding. For budding, 8 to 10 months old seedlings are used as rootstocks. Shield or ring budding is done during spring season with the swollen buds for optimum success. Generally seedlings or rooted cuttings are used as rootstocks.

Planting and after-care

Mulberry should be planted in square system at 6 x 6 m. Pits of 60 x 60 x 60 cm should be prepared during hot summer well in advance. After keeping them exposed for 15 to 20 days, they are filled up with a mixture of soil and well rotten farm yard manure (FYM) in 2:1 ratio. In sodic/rocky soils, the hard pan should be broken and 5 to 10 kg gypsum or 4 to 8 kg pyrite should be added or replacement of sodic soil with pond sediment along with 40 to 60 kg well rotten farm yard manure for better plant stand is recommended.

In India, planting season varies in different parts of the country. In Kashmir valley, planting in spring and late autumn is popular due to severe winter. In West Bengal, cuttings are planted during October-

November *i.e.,* late autumn. In Karnataka, the ideal season for planting is July-August *i.e.,* the onset of south-west monsoon. July and August are favourable months for planting of mulberry in Andhra Pradesh and Tamil Nadu. In most parts of north India, monsoon (July-August) or spring season (February-March) is the ideal time for planting. However, planting during extremes of winter and summer should be avoided.

In temperate regions, where sunshine is scarce, mulberry rows should have a north-south alignment. In lands with undulating topography, mulberry rows run parallel to the contour lines. In Karnataka and West Bengal, there is a practice to plant the cuttings directly in the field. Sometimes, when the field is not ready for planting, it becomes necessary to preserve the cuttings in sand beds under shade for about a week before planting. This method of storage helps in formation of "callus" and such cuttings give better establishment. In some cases, the cuttings are planted especially in prepared nursery beds with a close spacing of 6 to 8 cm and transplanted after a couple of months. This method has an advantage over direct planting as it reduces the incidence of failed pits in the mulberry plantation.

Manures and fertilizers applied in mulberry plantations have been worked out for maximum leaf production required for rearing silkworm. For fruit production, there is a need to work out doses of organic manures and chemical fertilizers for higher and quality production of mulberry.

Regular application of farm yard manure helps in improving organic matter content which improves microbial activity and water holding capacity of soil. In Karnataka, under rainfed conditions, application of about 10 tonnes of farm yard manure or compost per ha per year is recommended. Under irrigated conditions the dose should be 20 tonnes per ha per year. In West Bengal also, about 20 tonnes of farm yard manure/compost per ha per year is recommended.

As regards chemical fertilizers, application of balanced doses of fertilizer improves yield and quality of mulberry. The quantity of fertilizers to be applied depends upon the soil type, rainfall distribution pattern, spacing and pruning. Srivastava (1984) recommends 10 to 20 kg of FYM, 20 g nitrogen, 15 g phosphorus and 20 g potassium at the time of planting. A dose of 25 kg FYM, 40 g nitrogen and 30 g phosphorus per year should be applied till the age of 2 to 3 years, while of 30 kg FYM, 50 g nitrogen, 40 g phosphorus and 40 g potassium per plant per year is recommended for fully grown up trees. Care should be taken to apply fertilizers when there is sufficient moisture in the soil. Full doses of nitrogen, phosphorus and potassium are applied during February-March.

If rainfall is 800 mm or more and well distributed throughout the year uniformly, mulberry requires no supplementary irrigation. But such ideal conditions do not exist in the major mulberry growing states of the country. In such situations, efforts should be made to work out proper irrigation schedule for mulberry. Generally, basin irrigation at fortnightly intervals from February to April-May is sufficient. Drip irrigation is the most appropriate and input efficient system which is gaining popularity in areas of water scarcity. Drip irrigation ensures uniform distribution of water to plants, prevents water losses through conveyance and seepage, reduces weed growth and keeps the harmful salts below the root zone and saves the labour. Irrigations should be started after application of manures and fertilizers. No irrigation should be given during dormancy period in winter. Drip irrigation at 40 to 60 per cent of pan evaporation using four to six drippers per plant during fruit setting to maturity (February to May) is a general rule for optimum growth and fruit yield of mulberry.

Mulching the plant basin with organic waste materials or black polythene sheets reduces the water loss by evaporation. It also reduces weed population and improves the microbial activity of soil by improving the environment in the root zone. Regular use of organic mulches is helpful in improving the organic matter content which, in turn, improves water holding capacity of soil. Some green manure crops like sunhemp (*Crotolaria juncea*) and dhaincha (*Sesbania aegyptiaca*) sown in between the rows of mulberry plants can be conveniently pulled out later and used as mulching materials. Paddy straw, leaf mould, dry grasses, stubbles, etc. also serve as good mulch materials.

Weeding is very essential as these compete with mulberry plants for nutrient and water. Inter-cultivation can be done manually, bullock drawn or mechanized tillers. First weeding should be done 45 to 50 days after planting and repeated periodically during the first year of plantation. Thereafter, ploughing of inter-row spaces, periodically to a depth of about 15 cm should be done to eradicate weeds. Power tillers can conveniently be used in wider spacing plantations for inter-cultivation from the second year onwards.

Mulberry plants need to be periodically pruned to keep the production of fruits at the optimum level and maintain a low and manageable tree canopy during the productive phase. The best time for pruning in north India is during winter months when the tree remains dormant. In Karnataka and West Bengal, the practice in vogue is to raise mulberry as bushes only. The periodical pruning helps maintain the bush form for mulberry.

Harvesting and yield

Mulberry starts flowering and fruiting after 2 to 3 years. Flowering starts in February-March and fruit ripens during March-April. Moisture stress during February to May results in poor fruit yield. Colour break is the main index of fruit maturity in mulberry. In general, the fruits turn to black, red and white in *M. nigra, M. rubra* and *M. alba,* respectively at the time of maturity with attractive shining surface and typical aroma.

Time and stage of fruit harvesting depends upon the purpose whether fruits are to be used for fresh consumption or processing. Fruits are perishable in nature and stage of harvesting affects the shelf-life of fruits. Fruits should preferably be harvested in the early morning, packed in baskets with newspaper lining and sent to markets for sale. When markets are distantly located, harvesting can also be done in the late evening. Harvesting is done by shaking individual secondary branches of the tree at maturity or picking of individual fruits. There is severe damage to fruits if they fall directly on the ground. Therefore, fruits are allowed to fall on a cloth or plastic spread sheets held by two to three persons in suspended condition under the tree. The berries are very delicate and should be handled carefully. The mature fruits should be collected and disposed off quickly. Fruit yield depends upon variety and management practices. On an average, a full grown tree can produce 30 to 40 kg fresh fruit per tree with optimum management conditions.

References

Ottman, Yvonne (1987) *Fruit varieties* J., **41**: 4-7.

Rao, L. S. P., Rao, T. P. and Narayanan, E. S.(1963) *Curr. Sci.,* **32**:348-49.

Shivashankara, K.S., Roy, T.K., Rao, V.K. and Jalikop, S.H.(2006) Book of Abstracts, Nat. Symp. on Under-utilized Horticultural Crops held at IIHR, Bangalore on June 8-9, 2006, p.55.

Shukla, S.K. (2008) *CISH News,* **9**(2) : 1.

Singh, I. S. (2001) *Indian J. Hort.*, **58** (1&2) : 178-182.

Srivastava, R. P. (1984) Nutrition of Fruit Trees (In Hindi). ICAR, New Delhi, pp. 276-277.

Srivastava, R.P. and Kumar, S. (1994) Fruit and Vegetable Preservation - Principles and Practices, IBDC, Lucknonw, p. 264.

Tipton, J. (1994) *Journal of Arboriculture*, **20**(3) : 151-155.

Chapter 16 Karonda

Botanical Name	:	*Carissa carandas* L.
Regional/ Vernacular Names	:	Karamarda, avighna, karonda, karounda, karamcha, karavanda, karmarda, vaka, kalakkay, karekayi
Family	:	Apocynaceae

Karonda is a dichotomously branched large shrub or small tree, with strong, simple or forked thorns in pairs. It is, very often planted as live fence around field or orchard boundaries for protection and also for its fruits. The fruit, in clusters of 3 to 10 somewhat resembling large purple grapes, is oblong, broad-ovoid or round, 1.25 to 2.5 cm long; has fairly thin but tough, purplish-red skin turning dark-purple or nearly black when ripe; smooth, glossy; enclosing very acidic to fairly sweet, often bitter, juicy, red or pink, juicy pulp, exuding flecks of latex. Small fragrant white flowers are produced from early spring and the clusters of small, purplish to black fruits ripen from May to October.

Origin and distribution

Karonda is native to India, the coastal region of Natal, South Africa and is cultivated far inland in the Transvaal. It occurs throughout India, Myanmar, Sri Lanka, South Africa and Malaya, mostly on sandy or rocky soil in a wild state. In India, it is said to be wild in Bihar, West Bengal and south India and is widely planted as a protective hedge throughout India. Fruit is a more or less welcomed as by-product.

Nutritive, medicinal and other uses

The ripe fruits of karonda are sweet, edible and particularly suitable for tarts, puddings and jellies. The unripe fruit is sour and astringent, and used for pickles. It is said to possess anti-scorbutic properties. The fruits must be fully ripe, dark-red and slightly soft to the touch to be eaten raw. It is enjoyed whole, without peeling or seeding, out-of-hand.

Halved or quartered and seeded, it is suitable for fruit salads, adding to gelatins and using as topping for cakes, puddings and ice cream. Fruits can also be cooked to a sauce or used in pies and tarts. Stewing or boiling causes the latex to leave the fruit and adhere to the pot (which must not be aluminium), but this can be easily removed by rubbing with cooking oil.

Mature karonda fruits are used for making pickle, chutney, preserve and candy, while ripe fruits of maroon coloured variety may be processed into the syrup and jelly of excellent quality (Singh, 2001). Karonda fruits may also be dried for preparing condiments. Fruits are preserved whole by pricking, cooking briefly in sugar syrup and sterilizing in jars. Peeled or unpeeled, they are made into jam, preserves, syrup or sweet pickles. Jelly is made from slightly under-ripe fruits or a combination of ripe and unripe to enhance the colour. Kumar and Singh (1993) studied four types of karonda fruits from eastern part of Uttar Pradesh and found variability in fruit weight (1.6 to 4.7 g), number of seeds per fruit (5 to 11), moisture (87.5 to 90.0%), total soluble soilds (3 to 4.5%), acidity (0.65 to 5.99%), ascorbic acid (10.26 to 17.94 mg per 100 g), total minerals (0.8 to 1.13%) and iron (3.33 to 4.95 mg per 100 g) contents. The analysis of karonda fruits gave 18.2% moisture, 2.3% protein, 67.1% carbohydrates, 9.6% fat and 2.8% mineral matter (Anon, 1950). Karonda fruits contain 87 to 90 per cent pulp, 13.0 to 14.5% total soluble solids and 4 to 6% acidity (Singh, 2001).

The wood is white, hard and smooth, and is used for making spoons and combs. The root is a bitter stomachic and anthelmintic. A decoction of the leaves is given in the early stages of remittent fevers. Karonda is one of the trees on the leaves of which the tusser silkworm feeds. The plant has dense branches and sharp spreading thorns, and is suitable for fences.

Soil and climatic requirements

Karonda can be grown on a wide range of soils, including poor soils of arid and semi-arid regions. However, in good soils its growth and yields are better. The shrub thrives in dry, rocky terrain in Hawaii; in red clay or sandy loam in California, and in sandy or alkaline soils in Florida, though the latter may induce deficiencies of trace elements. The plant has moderate drought tolerance and high resistance to soil salinity and salt spray. It cannot stand water logging.

Karonda is very hardy and grows successfully under tropical and subtropical climatic conditions. It thrives throughout the state of Florida and enduring temperatures as low as 25°F (-3.89°C) when well-established. Young plants need protection when the temperature drops below 29°F (-1.67°C). Heavy rainfall and waterlogged conditions are harmful.

Cultivars

In general, the available karonda types are classified according to their fruit colour as green, pink and white. Due to seed propagation, considerable variation is observed in the fruit characters, such as shape, size, pulp quantity, quality and yield. Diversity occurs in the north-western India, particularly in Mount Abu, western ghats, in Khandala and Maharashtra. This offers great scope for improvement through seedling selections. In the recent past, karonda cultivars, *viz.* Pant Manohar, Plant Sudarshan and Pant Suvarna, have been released by G.B. Pant University of Agriculture and Technology (GBPUAT), Pantnagar for commercial cultivation. The chief characteristics of these cultivars are described below (Mishra, 2007):

Pant Manohar

With a leaf size of 2.35 × 1.36 cm, leaf area 2.83 cm^2, 44.7 leaves per shoot, its plants are medium-sized dense bushes, leaving fruits of 2.13 x 1.69 cm, dark pink blush on white background. Average fruit weight 3.49 g, seeds 3.94 per fruit, flesh 88.27%, dry weight 12.77%, TSS 3.92%, total titratable acidity 1.82% and yield 27 kg per plant.

Pant Sudarshan

With a leaf size of 2.475 x 1.447 cm, leaf area 2.84 cm^2, 41 leaves per shoot, size 2.16 x 1.69 cm, pink blush on white background. On ripening fruits become dark brown. Average fruit weigh 3.46 g, seeds 4.68 per fruit, flesh 88.47%, dry weight 11.83%, TSS 3.45%, total titratable acidity 1.89% and yield 29 kg per plant.

Pant Suvarna

With a leaf size of 2.41 x 1.41 cm, leaf area 2.8 cm^2, 45 leaves per shoot. Plants are upright growing and sparse. Fruit size 2.25 x 1.6 cm, dark brown blush on green background. Average fruit weight 3.62 g, seeds 5.89 per fruit, flesh 88.27%, dry weight 12.39%, TSS 3.836%, total titratable acidity 2.30% and yield 22 kg per plant. On ripening, fruit colour changes to dark brown.

At Rahuri in Maharashtra, some promising types No. 13, 16, 12 and 3 have been identified. Karonda Accession-3 has been released by

Regional Fruit Research Station, Vengurle for commercial cultivation, as it gives maximum yield, has large fruit size, higher fruit weight (16.23 g per fruit), better pulp content (92%) and low acidity (0.21%) (Salvi *et al.*, 2006)

Propagation

Karonda is propagated by seeds. The seedlings are slow in growth and ordinarily take one year to be ready for planting. Fresh seeds are sown in nursery in August or September for raising seedlings, since seeds usually lose their viability in 4 to 5 weeks. The seeds start germinating 40 to 45 days after sowing and complete in 60 days. Seedlings are raised in polythene bags during rainy season. Seedlings become ready in a year for transplanting. Bankar (1987) obtained enhanced germination (68%) with increased vigour of seedlings when seeds were soaked in 100 ppm GA_3, for 24 hours. GA_3 (1000 ppm) may be sprayed six times at monthly intervals for promoting growth of young karonda seedlings (Mishra, 2007).

Propagation of karonda by vegetative methods is not common. But, it is required for multiplication of improved cultivars in order to get true to type plants. Attempts have been made to propagate it by cutting, air layering, budding and soft wood grafting.

It is a difficult-to-root plant species, therefore, rooting does not occur in cuttings easily without using growth regulators. Attempts were made by Mehta and Singh (1960) to propagate karonda by stem cuttings. Cuttings from young plants and hedge sources proved to be the best for rooting and shoot growth. Divekar (1984) reported more rooting with 1000 ppm IBA treatment as quick dip. The cuttings are usually taken from mature, about one-year old shoot and should be 30 to 35 cm long with six leaves. Rainy season was observed to be the best period for propagation by cuttings. Treatment with 3000 ppm IBA and keeping the cuttings under mist markedly improved root formation. After 6 to 8 months, the plants attained transplantable height. Soft wood cuttings with 8000 ppm IBA in lanolin paste planted under mist in polyhouse conditions also gives a good rooting in karonda cv. Pant Sudarshan (Mishra, 2007).

A high percentage of success with air layering was recorded by Jauhari and Nigam (1958), when layers were treated with mixture of 7500 ppm each of IBA and NAA. Mishra and Singh (1990) reported that all the air layered shoots treated with IBA and NAA at 2500 to 10000 ppm in lanolin paste rooted, whereas untreated layers failed to root. The best season for air layering is the beginning of monsoon *i.e.,* June-July though it can be extended further up to September and the air layers should be treated with 400 ppm IBA to get good results (Divekar, 1984). Rooted layers were obtained after 3½ months of layering. Air layering with 5000 ppm IBA and stooling with 7500/10000 ppm IBA gives very good rooting in air and stool layers of white-pink strain of karonda (Mishra, 2007).

Propagation of karonda by shield budding is economic and convenient. Promising types of karonda could be budded on one-year old rootstocks of *C. carandas* (Bankar *et al.,* 1994). The best time for budding is July to August in low rainfall areas. Propagation of karonda by inarching also proved successful but when compared with other methods, the method is not very popular (Bankar *et al.,* 1994). Soft wood grafting is also successful in karonda and is a useful technique for grafting *in-situ* in arid regions.

Planting and after-care

The field is cleaned, thoroughly ploughed up and levelled. The pits of 30 to 45 cm^3 are dug up during summer and kept exposed for 15 to 20 days. The pits should be filled with a mixture of top soil and well-rotten farm yard manure (FYM) in a 3:1 ratio and then irrigated so that soil settles down. The pits should be brought to level of soil surface by filling additional soil. At the time of planting, care should be taken that plants are set in the centre of pits at the same level as they are in nursery. Irrigation should be done immediately after planting during July–August, but under irrigated conditions, planting can also be done in February-March. Orchards can be laid out in square, rectangular or hedge-row system depending upon the choice of orchardists and type of implements to be used. The planting distance from 3 to 5 m is kept both between row to row and plant to plant. For hedge-row planting plant to plant distance is kept 60 cm in two adjacent rows, whereas rows are spaced 2 m apart along the boundary of the orchard. The entire orchard can also be planted in hedge-row system in the same way and distance between 2 pairs of rows is kept 5 m.

As regards, nutrition of karonda plants, scientific information is scanty. But, it has been observed that application of 10 to 15 kg FYM per bearing plant before monsoon gives good results. A dose of 100 g urea, 150 g single super phosphate (SSP) and 75 g muriate of potash (MOP) is recommended by Mishra (2007) from G.B. Pant University of Agriculture and Technology (GBPUAT), Pantnagar for one year old karonda plant. The doses are increased in equal quantities every year up to the age of three years, stabilizing at 300 g urea, 450 g SSP and 225 g MOP per plant per year after three years of age.

In the beginning, plant requires plenty of nitrogen for its growth. It should be fertilized at bimonthly interval after dividing whole fertilizer requirement, in 6 split doses. Phosphorus and potassic fertilizers should be used after soil testing. In most cases, it is advisable to supplement the fertilizers with micronutrient sprays every year.

Karonda is a slow growing plant. Once established, it does not need much water. Moreover, it behaves as a xerophytic plant and, therefore, water requirement is very low. However, efforts should be made to collect run-off water during rainy season in root zone through land shaping or water harvesting methods. In young plants, light and frequent irrigations during summer are useful.

At the time of planting, each plant is supported with the help of wooden stake, so that it remains erect. Unwanted laterals are removed from time to time. Bearing trees usually do not require pruning. However, to give desired shape and making the stem clear, additional twigs are pruned. Diseases or broken twigs are also removed. Old hedge may be headed back to induce new growth.

Harvesting and yield

Harvesting in karonda starts 2 to 3 years after planting. The plants start flowering from February and continue till July. The mature fruits of early setting under irrigated conditions are ready in second week of June. Harvesting can be done from June to September. Both unripe and ripe fruits are harvested. Complete harvesting of all the fruits may not be done at one time. It is done two to three times. Green unripe fruits are available from March to May which are used for preparation of chutney and pickles, while ripe fruits are harvested in May-June. Colour change is a good indicator of fruit maturity *i.e.*, the change of fruit colour to dark purplish black. Pawar (1988) advocated that weight, volume, specific gravity, length, diameter and colour of the fruit can be considered as the physical indices, whereas moisture, TSS, sugars, acidity, pH and tannins as chemical indices of maturity in karonda.

Fruit yield varies with the soil, climate and management of plants. About 4 to 5 kg fruits per bush are obtained. Just after harvesting, fruits are kept in shade. Undesirable fruits are sorted out and rest packed in shallow baskets cushioned with leaves or straw and marketed. A fully grown up karonda plant may produce 25 kg fruits, on an average, under optimum conditions.

Off-season fruiting in karonda can take place in winter months in north India if grown under protected conditions, *i.e.* under polythene cover. Thus, fruits can be made available for processing throughout the year (Mishra, 2007).

References

Anonymous (1950) In : The Wealth of India : A Dictionary of Indian Raw Materials and Industrial Products, Vol. 3, CSIR, New Delhi, p.82.

Bankar, G. J. (1987) *Prog. Hort.,* **19** : 90-92.

Divekar, P.B. (1984) M.Sc. (Ag.) Thesis, KKV, Dapoli.

Jauhari, O.S. and Nigam, U.N. (1958) *Allahabad Fmr.,* **32** : 174.

Kumar, S. and Singh I.S. (1993) *South Indian Horticulture,* **41**(2): 108-110.

Mehta, K. C. and Singh, S.M. (1960) *Indian J. Hort.,* **17** : 31-37.

Mishra (2007) *Indian Hort.,* **52**(6) : 9-10.

Mishra, K.K. and Singh, R. (1990) *Annals Agril. Research,* **11** : 208-210.

Pawar, C.D. (1988) M.Sc. (Ag.) Thesis, KKV, Dapoli.

Salvi, B.R., Bendale, V.W., Kshirsagar, P.J., Malshe, K.V. and Chavan, S.S. (2006) Book of Abstracts, Nat. Symp. on Under-utilized Horticultural Crops held at IIHR, Bangalore on June 8-9, 2006, p. 33.

Singh, I.S. (2001) *Indian J. Hort.,* **58** (1&2) : 178-182.

Chapter 17 Cape gooseberry

Botanical Name	:	*Physalis peruviana* L.
Regional/ Vernacular Names	:	Tipari, tipariya, phopti, moti popti, buddabasara, tholthakkali, gudde hannu, rasbhary, mewar rashberi, chirput, chirboti, tankari, golden berry, makowi
Family	:	Solanaceae

Cape gooseberry is an annual fruit crop grown extensively in many of the tropical and subtropical regions of the world. The small fruits remain enclosed by the large thin papery epicalyx and resemble tomato in shape but with sweeter taste, better aroma and higher pectin content. However, in some regions with favourable climatic conditions, it may also be grown as a perennial crop. It is a herbaceous or soft-wooded plant, that usually reaches 0.6 to 0.9 m in height but occasionally may attain 1.80 m. The berry is globose, 1.25 to 2 cm wide, with smooth, glossy, orange-yellow skin and juicy pulp containing numerous very small yellowish seeds. When fully ripe, the fruit is sweet but with a pleasing grape-like tang. The husk is bitter and inedible.

Origin and distribution

Cape gooseberry is said to be originated in the western part of south America and is found to be native of Peru and northern part of Chile. The name cape gooseberry, most probably, was derived from the name of Cape of Good Hope of South Africa where it was commercially grown. The Portuguese and Spanish voyagers introduced the crop in various parts of the world. Presently, the species is commercially cultivated in tropical and subtropical regions of the world from Australia, Indonesia, Sri Lanka, India, Myanmar, Malaysia. Pacific islands, Bangladesh, southern China, Vietnam, Cambodia, southern Thailand, South African countries, Algeria and west African countries, Portugal, Italy, Greece and other Mediterranean and south European countries, Pakistan, northern part of south America and Florida in USA.

The crop is also said to be grown in the islands of Indian Ocean. However, in some of the north European countries, it is reported to be grown under cover or protection due to severity of the climatic conditions. The plant is widely grown in India for its edible fruits. The chief source of commercial supply is Uttar Pradesh, Rajasthan and Punjab. Other areas reported are Poona, Nilgiri hills, Coonoor and Araku valley (Andhra Pradesh).

Nutritive, medicinal and other uses

In nutritive value, cape gooseberry is comparable to banana, mango and papaya. In addition to being canned whole and preserved as jam, the cape gooseberry is made into sauce, used in pies, puddings, chutneys and ice cream, and eaten fresh in fruit salads and fruit cocktails. In Colombia, the fruits are stewed with honey and eaten as dessert. The British use the husk as a handle for dipping the fruit in icing. The edible portion constitutes 87% of the fruit. The food value of the fruit is given in Table 20 (Anon, 1969).

Table 20 : Composition of cape gooseberry fruit (per 100 g of edible portion)

Constituent	Content	Constituent	Content
Moisture (g)	82.9	Sodium (mg)	0.9
Protein (g)	1.8	Potassium (mg)	320
Fat (g)	0.2	Copper (mg)	0.19
Carbohydrates (g)	11.1	Sulphur (mg)	43
Fibre (g)	3.2	Carotene (I.U.)	2380
Mineral matter (g)	0.8	Thiamine (mg)	0.05
Calcium (mg)	11.1	Riboflavin (mg)	0.02
Magnesium (mg)	31	Nicotinic acid (mg)	0.3
Phosphorus (mg)	67	Ascorbic acid (mg)	49
Iron (mg)	2		

The ripe fruits are considered a good source of Vitamin P and are rich in pectin. The juice from the ripe fruits contains considerable quantity of pectin. The chief acid present in the fruit is citric acid but malic and tartaric acids are also present. Unripe fruits are poisonous. The plant is believed to have caused illness and death in cattle in Australia. In Colombia, the leaf decoction is taken as a diuretic and anti-asthmatic. In South Africa, the heated leaves are applied as poultices on inflammations and the leaf infusion is administered as an enema to relieve abdominal ailments in children.

Soil and climatic requirements

The soil and climatic requirements of the cape gooseberry are said to be similar to those of the tomato. It is said to thrive in varying types of soil ranging from sandy loam to laterite, but well drained sandy loam soils are considered to be the best. The crop prefers high and open lands with medium textured soil having adequate porosity for drainage. Low lying areas with poor drainage are not suitable for this crop. It can thrive on neutral as well as on acid soils. It is also seen to grow on saline soil in the coastal regions.

It is cultivated in both tropical and subtropical conditions. It tolerates mild cold up to 5^0C and higher temperature up to 35^0C, but frost (Philip and Khan, 1952) or excessive desiccation is unsuitable for the crop. As it cannot withstand frost, it must be given protection during winter in Punjab. The temperature around 21^0C is suitable for this crop. High humidity due to high rainfall increases the vegetative growth, disease incidence and makes the fruits insipid. It prefers drier climatic conditions for growth, pollination which is accomplished by insects, and development of quality fruits. It can be grown in plains as well as on the hills up to an elevation of 1800 m (Kuppuswami, 1954). It is grown as an annual in the plains of northern India and as a perennial fruit in the hills of the south.

Cultivars

There are more than 100 cultivars in cultivation in the world and about 60 of them are grown commercially. The red, white, yellow and green cultivars are in cultivation (Shanmugavelu, 1987).

Red cultivars

Bedford Red, Brown's Red, Champagne Red, Crown Lob, Conquering Hero, Speedwell, Warrington, Whinham's Industry, Wonderful

White cultivars

Antagonist, Freedom, White Eagle, White Lion, White Swan

Yellow cultivars

Bedford Yellow, Broom Girl, Golden Drop, Lady Houghton, Leveller No.l, Pilot

Green cultivars

Angler, Drill, Green Ocean, Green Overall, Matchless, Shiner

In India, there are no standard cultivars in cultivation. The available types can be classified into annual and perennial types or round and elongated types. Classification can also be done on the basis of colour. There is a need to identify, characterize and popularize superior types for promoting this commercially important fruit crop.

Propagation

The plant is widely grown from seed. There are 180 to 280 seeds per g and, since germination rate is low, 100 to 250 g seed is needed for planting in one ha area depending upon the germination percentage. The seeds of cape gooseberry are small and difficult to sow, therefore, the seeds are mixed with wood ash or pulverized soil for uniform sowing. Seed beds of 4 m length, 1 to 1.5 m width and 10 to 15 cm raised seedbeds are prepared for sowing. However, for small area, the seeds may be sown in seed pans for easy and efficient handling. Seeds after sowing in the seed bed, should be covered with leaf mould and watered. Seedbeds should be kept moistened till the seeds germinate.

The perennial crop which can thrive up to 4 years (Naik, 1949) may be propagated by vegetative means. For vegetative propagation, shoot cutting or layering is practiced with varying success. In these cases, one year old healthy plants are selected for the purpose of propagation. Cuttings measuring at least 10 to 25 cm long are selected from the mature but succulent shoots during the rainy season when the atmospheric temperature and humidity remains relatively high. Layering gave higher (80%) success than soft wood cuttings, which gave 20% success (Ramasomayazulu, 1954) and the method was considered suitable for commercial use, particularly for the perennial type. Etiolation helps in early regeneration of roots in the cuttings. The optimum temperatures for rooting of cape gooseberry cuttings were 18 to 30°C. The cuttings rooted best in 60 to 85 per cent day light. It can be propagated by cuttings planted under mist without using any growth substances. The cuttings should, however, be taken at right degree of maturity. Rooting was best in soft wood cuttings taken from the upper part of the shoots of the third and fourth tier irrespective of the time of taking. Cuttings taken from the upper part of the basal roots, rooted best late in the season, whereas those from the lower part of similar roots, rooted best early in the season.

Planting and after-care

Soil should be prepared thoroughly to a fine tilth before transplanting. The field should well drained and well manured. Time of seed sowing depends on the season of growing the crops. If desired to grow in open condition, the seeds are sown 6 to 7 months prior to planting. Seedlings are planted when they attain a height of 15 to 20 cm. Seedlings are transplanted in square system with a spacing of 0.8 to 1.0 m, both between row to row and plant to plant. The transplanted seedlings may require shade and light irrigation for a few days, if the soil moisture is not sufficient. Planting is done in September or October on heavily manured ridges, 45 to 90 cm wide in south India. In north India, the seeds are sown in May to July and the seedlings are planted when they are about 20 cm high.

At the time of planting, application of 20 tonnes per ha farm yard manure and application of 250 kg nitrogen, 240 kg phosphorus and 240 kg potassium per ha is recommended to get higher yield and quality fruits (Shanmugavelu, 1987). Plants receiving high nitrogen and potassium produced more and highly coloured fruits.

In some places, it has been reported to apply 20 tonnes of organic manure, 350 kg of single super phosphate, 120 kg of muriate of potash and 100 kg of urea per ha. Besides this, first top dressing of 75 kg of urea per ha at 30 day stage and second of 50 kg urea per ha at 50 to 55 day stage should be done. Spraying of 0.1% saline water twice up to the age of three months at an interval of 30 days, starting one month after the transplanting is in vogue to decrease the vegetative growth and increase the productivity of the crop in some parts of India (Chattopadhyay *et al.*, 1984). Singh *et al.* (1977) found that higher nitrogen in combination with phosphorus and potassium increased plant growth and yield. The quality of fruits, in terms of total soluble solids was also improved due to nitrogen application. The observations were confirmed by Prasad *et al.* (1985) where 100 kg of nitrogen, 60 kg of phosphorus and 60 kg potassium per ha produced highest yield and juice content of fruits.

The crop requires 7 to 8 irrigations, in the absence of rainfall. During drier periods, the crop requires irrigation at an interval of 10 to 15 days depending upon soil moisture. Irrigation during flowering helps in minimizing flower drop, while during fruit development, it improves the size of fruit and reduces fruit drop. However, care should be taken not to irrigate the crop excessively during fruit maturity which induces fruit splitting and spoilage.

Weeding by light hoeing or forking is a common practice in commercial cultivation. The practice of earthing up across the slope in undulating land supports the plant and facilitates irrigation. In the tropics, a weed, *Physalis minima* is often confused with cape gooseberry and creates problem which should be carefully removed from the field. Application of Alachlor, a herbicide @ 2.5 kg a.i. per ha as pre-emergence spray, 20 days prior to planting was found effective for control of weeds (Sarkar, 1992).

One year old gooseberry tolerated simazine @ 3 kg per ha applied five days after planting (Shanmugavelu, 1987). It gave effective weed control and resulted in good plant development and increased yield. Mulching with black polythene film as compared to hoeing and hand-weeding promotes growth to some extent, but cuts down expenditure on manual weeding, compensating the initial expenditure on mulching.

Harvesting and yield

Cape gooseberry fruits become harvestable 55 days after anthesis (Sarangi *et al.*, 1989). Immature fruits should not be harvested. Rather, mature and turning fruits as ascertained by increased TSS, total and reducing sugars, and decreased acidity are harvested for marketing (Sarangi, 1989). Maturity of berries is also indicated by change of colour of the epicalyx from green to pale brown or yellowish orange. Fallen berries are said to be more uniform in maturity than hand picked ones. As the fruits become ready for harvest, a number of harvests may be required to collect the properly ripe fruits. The fruits may be marketed directly or these may be stored at room temperature for a number of days for ripening which is ascertained by change of colour from green to golden yellow (Pradhan, 1989). Fruits with intact epicalyx may be best stored in brown paper packets, for longer period at room temperature (Sarkar *et al.*,1993).

Cape gooseberry yield often varies with type and cultural management adopted for production. The yield of fruits per plant is said to be about 2 to 5 kg. On an average, the yield of cape gooseberry fruits varies from 4.5 to 6.0 tonnes per ha. The perennial variety was reported to yield higher (Naik, 1949). In south India, first picking lasts from January to April; then the bushes put forth new blossoms in May and a second picking is possible from July to September. The plants are kept only for two years. However, it is customary to keep them for four years at Coonoor, where yields are more than 30 tonnes per ha. In north India, the fruits are available from January to April, peak period being February-March.

References

Anonymous (1969) The Wealth of India-Raw Materials, Vol. VIII, CSIR, New Delhi, pp. 38-40.

Chattopadhyay, T.K., Mazumdar, B.C. and Nandi, A.K. (1984) Scientific Cultivation of Fruits, Central Book Publishers, Kolkata, pp.159-161.

Kuppuswami, B.S. (1954) *South Indian Horticulture*, **3**(4) : 104-107.

Naik, K.C. (1949) South Indian Fruits and their Culture, P. Varadachary & Co., Chennai.

Philip, A. G. and Khan, M.A. (1952) *Punjab Fruit Journal*, **15**(53) : 17.

Pradhan, U. M. (1989) M.Sc. (Ag.) Thesis, BCKV, Kalyani.

Prasad, I.D., Sengupta, B.N., Singh, R.K. and Singh, S.P. (1985) *Haryana J. Hort. Sci.*, **44**(3-4) : 151-155.

Ramasomayazulu, M.V. (1954) *South Indian Horticulture*, **1**(1) : 145-47.

Sarangi, D.K. (1989) M.Sc. (Ag.) Thesis, BCKV, Kalyani.

Sarangi, D.K., Sarkar, T.K., Roy, A.K., Jana, S.C. and Chattopadhyay, T.K.(1989) *Prog. Hort.*, **21**(3-4) : 225-228.

Sarkar, T. K. (1992) Ph. D. Thesis, BCKV, Kalyani.

Sarkar, T. K., Pradhan, U.M. and Chattopadhyay, T.K. (1993) *Annual Agril. Res.*, **14** (4) : 396-401.

Shanmugavelu, K.G. (1987) In: Production Technology of Fruit Crops, SBA Publications, Kolkata, pp.433-437.

Singh, U.R., Pandey, I.C. and Prasad, S.R. (1977) *Punjab Hort. J.*, **17**(3-4) : 148-151.

Chapter 18 Khirni

Botanical Name	:	*Manilkara hexandra* (Roxb.) Dubard syn. *Mimusops hexandra* Roxb.
Regional/ Vernacular Names	:	Khirni, khirkhejur, ranjana, rayan, raini, manjipala, pala, palla, palai, bakula, khiri, khirakuli
Family	:	Sapotaceae

Khirni grows naturally or planted along road sides almost throughout India both for its ornamental value and also for its sweet edible fruits. It is one of the popular trees found growing in forest and wastelands. The fresh fruits are eaten by the local people. Tree is medium to large in size, evergreen with spreading grown and straight massive bole. The tree is ideally suited for growing on waste and barren lands. The khirni is commercially used as rootstocks for sapota. Ripe fruits are eaten fresh or dried. Trees of very big size are common in Pratapgarh area of Rajasthan along with *Madhuca latifolia* (Sen, 2001).

Origin and distribution

Khirni is a native of tropical south-east Asia. It occurs naturally in forests and common lands particularly central and peninsular India, in particular. Chanderi in Ashok Nagar (previously known as Guna) district of Madhya Pradesh is famous for availability of large number of naturally occurring khirni trees and production of best quality fruits. Fruits from this area are transported to Mumbai and other cities and fetch an attractive price.

Nutritive, medicinal and other uses

Small yellow olive like fruits are sweet when fully ripe. They are eaten fresh or dried. The young fruits are boiled and eaten. The seeds, fried or roasted, taste like peanuts but contain an alkaloid and are not to be used in excess. The fruits weighing 1.76 g to 4.44 g each, contain pulp

(82.12 to 90.33%) and TSS (21.8 to 29.3°Brix). Nutritional composition of fruit is given in the Table 21 (Anon, 1957).

Table 21 : Composition of khirni fruits (per 100 g of edible portion)

Constituent	Content	Constituent	Content
Moisture (%)	68.61	Iron (mg)	0.92
Protein (%)	0.48	Carotene (IU)	675
Fat (%)	2.42	Thiamine (µg)	70.33
Carbohydrates (%)	27.74	Riboflavin (µg)	77.41
Mineral matter (%)	0.75	Nicotinic acid (mg)	0.66
Calcium (mg)	83	Ascorbic acid (mg)	15.67
Phosphorus (mg)	17		

The seed contains 24.6% of edible oil known as Ryan oil. The oil contains palmitic acid (18.9%), stearic acid (14.1%), oleic acid (63.2%) and linoleic acid (2.7%). The seed contains a bitter saponin, which is left in the cake after extraction of oil, a sapogenin bassic acid. The tree yields a gum, the bark contains 10% tannins and may be used for tanning purposes. It is used as a rootstock for sapota/cheeku. Leaves are used as cattle fodder. The tree yields a gum. The bark contains 10% tannin and may be used for tanning purposes. It is used in fevers and as general tonic. The bark retards the fermentation of toddy. The seed oil is considered demulcent and emollient.

Soil and climatic requirements

Khirni is best suited for sandy wastelands of semi-arid regions. It grows well on all types of soils. However, well drained loamy soils and medium black soils are ideal for its cultivation. Soils with hard rocky substratum or impervious clay are not suitable. The tree is hardy and grows well in tropical and subtropical regions of India.

Cultivars

There are no specific cultivars recognised so far in khirni. However, good amount of variability occurs in size, shape, colour, pulp content and taste of the fruits. There is a need to screen the available genotypes and select the elite clones for commercialization as the fruit has very good market value.

Propagation

Khirni is raised from seed. The seedlings are often raised *in-situ*. This can be achieved, to a great extent, by raising the plants in polythene bags. Seeds are extracted from fresh fruits and are sown in rainy season for raising seedlings. Seed viability of freshly extrated seeds is more. Khirni seeds are viable up to five months. Fresh Khirni seeds need to be scarified using concentrated H_2SO_4 for two minutes followed by washing in running water and soaking in 1000 ppm thiourea for six hours before sowing in seed beds (Singh *et al.*, 2007). Khirni seedlings are transplanted at four leaf stage. The seedlings take about 7 months to attain graftable size and thickness.

Various vegetative methods of propagation have been attempted at Central Institute for Subtropical Horticulture (CISH), Lucknow. Of them, veneer grafting in July-August was found most successful (Singh *et al.* 2007). The seedlings were raised *in-situ*. Soft wood grafting has also given good success in some parts of the country.

Planting and after-care

One to two year old seedlings are suitable for transplanting in field as well as for using as rootstocks for sapota. It is planted at a distance of 8 x 8 m. Pits of 1 x 1 x 1 m size are dug and filled with 20 to 30 kg well rotten farm yard manure mixed with top soil of the pit. Half kg of di ammonium phosphate (DAP) should also be mixed with farm yard manure. About 20 kg of well-rotten farm yard manure should be applied per plant per year. For bearing trees, this dose should be increased to 50 to 80 kg depending upon the size and age of tree. This is sufficient for growth, flowering and fruiting. Manuring in bearing trees, having luxuriant growth should be avoided. Adoption of effective water management practices is essential. These include water harvesting techniques, conservation of soil and plant water and use of protective irrigations during establishment of orchard and drought periods. Improper drainage and aeration of tap roots results in decline of orchards. Water harvesting techniques should be adopted to conserve rain water. Irrigations should be given during critical periods such as during initial years of orchard establishment and also during drought periods.

Harvesting and yield

Flowering takes place from November to January and the fruits ripen from April to June. Ripe fruits when turn to golden yellow colour are harvested. A fully grown up seedling tree can produce 100 to 300 kg fruits every year.

References

Anonymous (1957) Report Department of Nutrition, Govt. of Mumbai, p.26.

Sen, L.N. (2001) Proc. Winter school on exploitation of Underutilized Fruits held at MP University of Agriculture and Technology, Udaipur from Oct. 3-23, 2001, pp. 198-199.

Singh, A.K., Shukla, S.K., Bajpai, A., Singh, A. and Singh, M.P. (2007) *Indian Hort.*, **52**(5) : 4-5.

Chapter 19 Barhal

Botanical Name	:	*Artocarpus lakoocha* Roxb.
Regional/ Vernacular Names	:	Monkey jack, dephal, wotomba, kamma-regu, vatehuli, myankdok, lakuch, lakuchi, badahar, tampang, lokhat
Family	:	Moraceae

Barhal is a valuable tree species of India used for fruit, furniture, timber and fodder. The tree is deciduous and grows 6 to 9 m tall with large and leathery leaves. The tree is cultivated for its edible fruits, which are roundish, 5 to 10 cm in diameter, ill-shaped and of yellow, orange or dirty yellow colour with sweetish or sour taste.

Origin and distribution

Barhal is native to humid sub-Himalayan regions of India. It occurs in the sub-Himalayan tract from Kumaon eastwards to Assam and southwards to Myanmar and the Andamans. It is naturally available in Bihar, Orissa, West Bengal and central states. In the west coast, it is found from Konkan southwards to Malabar and Travancore.

Nutritive, medicinal and other uses

The ripe fruits are generally eaten fresh. Raw fruits and male flower spikes are acidic and astringent and are utilized for pickles and chutney. Mature fruits may also be processed into pickle and jelly, while ripe fruits are suitable for ready-to-serve beverage and squash (Singh, 2001). As regards nutritive value of fruits, Singh *et al.* (1999) recorded average fruit weight of barhal from 135.2 to 208 g, pulp 49.8 to 72%, seed 12.4 to 21.15%, TSS 17.2 to 21.2%, acidity 0.7 to 1.1%, pectin 0.6 to 1.0% and minerals 0.8 to 1.1%. Vitamin C from 6.2 to 18.3 mg and vitamin A from 104 to 150 IU per 100 g of edible portion of fruit was also recorded.

Raw fruits are known to have 82 g moisture, 1 g each of protein, fat and minerals, 2 g fibre, 13 g carbohydrates, 50 mg calcium and 20 mg phosphorus in 100 g of raw barhal fruit. Each fruit contains 20 to 30 seeds that are fleshy with thin seed coat. The tree is also valued for feed and timber. It is an important fodder tree of Nepal (Kessler, 1981). The hard wood of barhal, is comparable to famous teak wood. It is straight or interlocked grained, very coarse and even textured and is durable outdoors as well as under water. It is used for construction, furniture, boat making, and cabinet work. Tree bark containing 8.5% tannin and is chewed like betel nut. It is also used to treat skin ailments. It yields a durable fibre good for cordage. The edible fruit pulp is believed to act as a tonic for the liver. The wood and roots yield a lavish colour dye. Barhal seeds and milky latex are purgative. Seeds contain artocarpins (ALA I and ALA II), the isolectins which exhibit high haemagglutination activity (Wongkham, *et al.*, 1995). The haemagglutination activity of ALA was demonstrated in various organs of the plant except fruit flesh. The highest activity was found in the seeds (14400 units per g fresh weight), while it was the lowest in leaves (5 units per g fresh weight).

Soil and climatic requirements

Barhal grows on a wide variety of soils, though it prefers a rich, well drained, deep alluvial soil. It can be cultivated on open textured light soil or laterite soil provided nutrients are available.

Barhal can be grown successfully in both tropical and subtropical climatic conditions. It flourishes in humid hill slopes also up to an elevation of 1200 m. Quality of fruits deteriorates at higher altitudes. It also grows well in arid, semi-arid and warmer plains of south India. Cold weather and frost are harmful.

Cultivars

Barhal trees have lot of variability in terms of fruit characters. Barhal fruits were grouped into two categories (Singh *et al.*, 1999) based on colour of fruit pulp, one with deep orange pulp and other with creamy yellow pulp. They recorded that fruits with orange coloured pulp had higher content of pectin, vitamin and minerals when compared with fruits having yellow pulp. Ready to serve and squash prepared from fruits having orange colour pulp scored better than the fruits with yellow pulp.

Ten barhal genotypes identified through explorations in Uttar Pradesh were characterized for various physico-chemical traits (Shukla *et al.*, 2008) and the fruits of different genotypes showed considerable variation with respect to fruit characters. Average fruit weight from 97.0 to 460 g, fruit length from 5.48 to 9.45 cm, fruit breadth from 6.49 to 10.04 cm, number of seeds from 26.0 to 127.5 per fruit were recorded. The biochemical parameters of fruits revealed total soluble solids (TSS) from 18.87 to 27.78^{0}Brix, acidity from 0.74 to 2.37 per cent, ascorbic acid from 13.67 to 24.54 mg per 100 g and total sugars from 12.60 to 17.82 per cent in various genotypes. Out of ten genotypes, four accessions, *viz.*, CISH Barhal-7, CISH Barhal-3, CISH Barhal-10 and CISH Barhal-6 were found most promising (Shukla *et al.*, 2008).

Propagation

There are a number of important problems associated with regeneration and propagation of barhal, (1) tree population of barchal is gradually decreasing due to poor seed viability and extensive exploitation of tree for food, timber, and other uses; (2) seeds, once extracted from the fruit, quickly lose viability within a week, or sometimes even in few days; and (3) vegetative propagation methods such as rooting of hard wood or soft wood stem cuttings have not been successful (Napier and Robbins, 1989).

Barhal is most commonly propagated by seed. Freshly extracted seeds should be sown, or seeds lose viability. Barhal can also be propagated by air layering and cuttings, but it is difficult to propagate by vegetative means.

Planting and after-care

For planting barhal, pits of 1 × 1 × 1 m size spaced at 10 × 10 m in square system of planting are dug at least a month prior to planting. The spacing can be increased to 12 × 12 m in fertile soil to avoid crowding of trees in later stages. After exposure of the soil to the sun for sufficient period, 30 to 40 kg well-rotten farm yard manure (FYM) and 500 g super phosphate are mixed with the top soil of each pit followed by filling of pits. Chlorpyriphos dust @ 100 g per pit may be mixed with the mixture to avoid termite attack. The best time for planting seedlings is July-August in north India. Prolonged dry weather after planting may lead to the death of plants. The tap root should not be disturbed while planting to avoid damage to plants.

The young plants should be adequately guarded initially for about two years. Hand watering of young plants during initial years is necessary for assured survival and good growth of plants. In colder regions, protection against frost at least during first few years is a must. Cleaning of basins by regular weeding/interculture and mulching in tree basins are good for proper maintenance of orchard. The barhal is not irrigated normally. Irrigation during dry periods is considered essential in arid regions for normal growth. In order to economise use of water, ring system may be adopted for irrigation. For young orchards, hand watering is necessary during first 2 to 3 years. The frequency of irrigation depends on the soil moisture condition.

Barhal trees need nutrition for regular and good cropping. Application of 5 kg well rotten FYM, 100 g nitrogen, 50 g phosphorus and 50 g potassium is sufficient for one year old barhal tree. This dose should be increased every year up to 10 years of tree age. After 10 years, the dose is stabilized and 50 kg FYM, 1 kg nitrogen, 0.5 kg phosphorus and 0.5 kg potassium should be applied every year. Full doses of manures and fertilizers should be applied in January-February. The manures and fertilizers should be spread in the basin and thoroughly mixed in soil. The fertilizers should be applied in an area under the tree canopy leaving 1 to 1.5 m around the tree trunk.

Harvesting and yield

Barhal seedling trees take five years to come into bearing. The yellow or orange-yellow male flowers and reddish female flowers of barhal are borne separately on the same trees. Fruits are nearly round or irregular, from 5 to 12 cm in diameter and have a velvety surface. The barhal fruits are dull-yellow with pink tinge and sweet-sour pulp. Fruit yield can be obtained up to 80 kg per tree with fruit weight ranging from 200 to 350 g.

References

Kessler, C.D.J. (1981) *International Tree Crops Journal*, **1** (40) : 245-272.

Napier, I. and Robbins, M. (1989) Forest Seeds and Nursery Practice in Nepal, pp. 94-95.

Shukla, S.K., Ali, A., Pandey, D. and Pandey, G. (2008) Souvenir - Nat. Sem. on Sustainable Horticultural Research in India : "Perspective, Priorities and Preparedness" held at BBAU, Lucknow on April 14-15, 2008, p. 20.

Singh, D. B., Attri, B.L., Sharma, T.V.R.S. and Sreekumar, P.V. (2001) *Journal Applied Hort.*, **3**(1): 60-62.

Singh, I.S. (2001) *Indian J. Hort.*, **58** (1&2) : 178-182.

Singh, I.S., Srivastava, A.K. and Singh,V. (1999) *Journal Applied Hort.*, **1** (1) : 34-37.

Wongkham, S., Wongkham, C., Boonsiri, P., Simasathiansophon, S., Trisonthi, C. and Atisook, K. (1995) *Phytochemistry*, **40** : 1331–1334.

Chapter 20 Amra

Botanical Name	:	*Spondias pinnata* (L. f) Kurz. syn. *S. mangifera* Willd., *S. acuminata* Roxb
Regional/ Vernacular Names	:	Wild mango, hogplum, amara, jangli aam, amna, amra, ambra, ram-amba, amb, anbada, adavimaamidi, ivurumaamidi, touramamidi, kotamara, katamara, mari-man-chedi, manpulichi, ambalam, ampallai, ambiram, eginam, ibangam, kathimagirangai, pullipullama, amatekaye, ambattemara, poondi, gooddamate, kaadamute, marahunsie, vrykshamla, kataambolamm ambazham, manpuli, puliman, ambula, amodo, amaratoko, amora, amratenga, bahamb, ambar, ambaru, arbaru, gue, ranchi ling
Family	:	Anacardiaceae

Amra (Ambar) is a small aromatic, deciduous tree growing up to 27 m high and 2.5 m in girth found wild or cultivated almost throughout India, ascending up to an altitude of 1500 m in the Himalayas. Flowers whitish or greenish white, polygamous, appear in large terminal panicles. Fruits drupaceous, yellow green or rich olive green, mottled with yellow and black, smooth, ovoid, 4 to 7 cm long, 2 to 5 seeds, one being perfect, oblong elliptic. Tree is often grown for fencing and shade.

Origin and distribution

The tree is native to tropical Asia, grows and fruits well in southern Florida. The tree is grown in West Indies, central and south America, south-east Asia, Malaysia and India. In India, tree is more common in central, north-eastern, south-west, Andamans and southern parts of India. In Uttar Pradesh, the tree is common in central and eastern districts.

Nutritive, medicinal and other uses

The fruit is used as a vegetable when green and as a fruit when ripe. It is used as a condiment and can be made into chutney, stews, pickles, jams and jellies. Cattle, deer, monkeys and squirrel greedily eat the fruits. The wood is employed for packing cases, tea chests for floats, canoes or boats. It is suitable for match splints but not for match boxes and for non-ornamental plywood. The wood is used for fuel, though of poor quality. The composition of fruit is given in Table 22 (Anon, 1969).

Table 22: Composition of amra fruits (per 100 g of edible portion)

Constituent	Content	Constituent	Content
Moisture (g)	90.3	Phosphorus (mg)	11
Protein (g)	0.7	Iron (mg)	3.9
Fat (g)	3.0	Vitamin A (IU)	450
Carbohydrates (g)	4.5	Thiamine (mg)	0.02
Fibre (g)	1.0	Riboflavin (mg)	0.02
Mineral matter (g)	0.5	Nicotinic acid (mg)	0.3
Calcium (mg)	36	Vitamin C (mg)	21

The fruit is astringent and anti-scorbutic. It is stated to be useful in bilious dyspepsia. It is eaten either green or dry as an antidote for wounds caused by poisonous arrows. The juice is applied in earache. The plant is reported to have anti-tubercular properties. The bark is aromatic, astringent and refrigerant. It is useful in dysentery and diarrhoea and is also given to prevent vomiting. A paste of it is used as an embrocation for both articular and muscular rheumatism. A decoction of the bark is given in gonorrhoea. The bark is used for tanning. The root is useful in regulating menstruation.

The tree exudes a mild, insipid, yellowish or reddish brown, stalactiform gum with smooth, shining surface, resembling gum arabic, except being darker. It forms a gelatinous mucilage with water but is not wholly soluble. It is used as demulcent and for the purpose of fumigation. The gum is sometimes employed as an adhesive.

The leaves are aromatic, acidic and astringent. They are used for flavouring. Their juice is applied in earache. The flowers are sour and eaten as such, made into a curry, or used as a flavouring.

Soil and climatic requirements

It grows on all types of soil as long as they are well drained, but a deep, well drained loamy soil is desirable. The tree flourishes in humid tropical and subtropical areas. The tree is distributed in deciduous and often dry forests, but it is nowhere abundant or gregarious in India, It grows in regions where the maximum temperature in shade ranges from 38 to 48^{0}C, the absolute minimum temperature is in between -1 and 16^{0}C and the rainfall is in between 75 and 375 cm or more. It tolerates low temperature to a great extent during winter as it sheds leaves during winter in most parts of north India.

Cultivars

There are no standard cultivars of amra as its cultivation is rare. Fruits from naturally occurring trees are generally sold in local markets. Generally, two types of amra trees are found depending upon the shape of fruits. One type has oval to oblong fruits, while another type has fruits of round shape. Some have acidic while others have sweetish taste and can be used as salad. Similarly, there is lot of variability in terms of stone size. There is a need to record the variability in amra and select promising genotypes for further commercialization.

Propagation

The plant is easily propagated by seed. Small heaps of stones with flesh removed are seen scattered on the forest floor during December to February along with the droppings of the deer. The seeds collected from such heaps germinate well. Germination decreases by 50% after a year of storage. Suitable clipping of seeds increases the germination percentage. Clipped seeds are put in polybags in December-January with a layer of hay. Regular watering is done since radicle gets dried up. Germination is noticed after 15 days. One year seedlings are planted in the field. Sowing the seeds directly in the field gives better results than transplanting, provided the stones are covered with earth. However, polythene or basket raised seedlings transplant better. Seedlings should preferably be grown in polythene bags, filled with a mixture of soil, sand and well rotten farm yard manure (FYM) in 2:1:1 ratio. Such seedlings are easily transplanted during ensuing rainy season.

The plant is also propagated by semi-hard/hard wood cuttings. Planting of stumps measuring 1.7 cm in diameter gives very good results.

Planting and after-care

Vegetatively propagated amra plants can be planted in square system of planting at 5 to 6 m spacing both ways, while seedling trees should be planted at 8 to 10 m. Pits of 60 × 60 × 60 cm are dug a month prior to planting. The pits are refilled with mixture of top soil and well rotten FYM in equal quantities. Following refilling, irrigation is done to ensure that soil in the pits settles down well before planting. The ideal time for planting is rainy season, but they can also be planted during February-March, provided irrigation facility is available. Seedlings or rooted cuttings are planted in the centre of pits. Irrigations are required at weekly intervals just after planting, in case there is no rain.

Systematic information is not available on nutrient doses to be applied for amra trees. But, 30 to 40 kg well rotten FYM per tree should be applied every year for optimum flowering and fruiting. In the initial stage, desired shoots are allowed to grow in different directions to form strong framework of desired shape. In later stages, it does not require any training or pruning. No irrigations are required during winter when the tree remains dormant, but irrigations at an interval of 15 to 25 days should be given during summers for proper growth and fruiting.

Harvesting and yield

Flowering occurs during spring season (March-April) in north India and the fruit ripens during December to February. Fruiting starts in about 3 to 5 years. A fully grown up tree produces 100 to 300 kg of fruits per year. Fruits are harvested both at raw and mature stage for pickle making. Unripe fruits are harvested during July-August, while mature fruits with hard endocarp harvested during winter are also used for pickle making. Ripe fruits harvested during winter are often used for raising seedlings.

References

Anonymous (1969) The Wealth of India- Raw Materials, CSIR, New Delhi, Vol. X, pp. 20-21.

Chapter 21 Star gooseberry

Botanical Name	:	*Cicca acida* (L.) Merril syn. *Phyllanthus acidus* (L.) Skeels
Regional/ Vernacular Names	:	Country gooseberry, lavaliphala, harfarouri, hariphal, rachayusirika, aranelli, kirinelli, star aonla, sour cherry, Tahitian gooseberry, Otaheite gooseberry
Family	:	Euphorbiaceae

Star gooseberry is a small fruit tree frequently cultivated in home gardens and backyards in almost all parts of India, chiefly for the fruits. The plant is a curious intermediary between shrub and tree, reaching 2 to 9 m in height. The tree is dense and bushy. The crown is composed of thick, tough main branches, at the end of which are clusters of deciduous, greenish, 15 to 30 cm long branchlets. The branchlets bear alternate leaves that are ovate or lanceolate in form, with short petioles and pointed ends. The leaves are 2 to 7.5 cm long and thin, they are green and smooth on the upperside and blue-green on the underside. The flowers can be male, female or hermaphrodite. They are small and pinkish and appear in clusters in 5 to 12.5 cm long panicles. Flowers are formed at leafless parts of the main branches, at the upper part of the tree. The fruits are numerous, oblate, with 6 to 8 ribs, develop so densely that they actually form spectacular masses. They are pale yellow or white, waxy, crisp and juicy, and very sour. It has only one seed in each fruit.

Origin and distribution

This tropical or subtropical species originated in Madagascar, then carried to the East Indies. Now, it is generally found in south-east Asian countries, such as southern Vietnam, Laos, Indonesia and northern Malaya. It also occurs in the Indian Ocean islands of Mauritius, Reunion and Rodrigues and also in Guam, Hawaii and several other Pacific islands. In 1793, the plant was introduced to Jamaica from Timor. From there, it progressively spread to the whole Caribbean region, as far as the Bahamas or Bermuda. It is now naturalized in central and south America. In the United States, the tree is occasionally found as a curiosity in Florida. Great natural diversity is available in north-eastern Hill region and southern parts of India. However, it is distributed throughout India.

Nutritive, medicinal and other uses

The fruit is sour, edible, astringent berry, which is eaten raw, cooked or pickled, or is made into jam, jelly. Analysis of fruits gave 89.63% moisture, 0.9% protein, 0.58% crude fibre, 7.29% carbohydrates, 0.84% ash and 1.7% acidity (Anon, 1950). The fruits are nutritive and are a rich source of vitamin C and minerals. They are divided into 6 to 8 segments and single seeded. It is rich in pectin, vitamin C and acid contents. The fruits are non-climacteric, hence attain no respiratory peak at maturity. They have good shelf-life and can be kept for 8 to 10 days without any deterioration in their quality. The fruits are ideal for chutney, preserve and pickle making. It is a heavy-bearer and has short fruit maturity period besides being tolerant to biotic and abiotic stresses. Many processed products, *viz.* preserve, candy, pickle, jelly, etc. can be prepared from the fruit commercially. Leaves are also used as vegetable. The physico-chemical characters of star gooseberry fruits are given in the Table 23 (Kundan Kishore, 2007).

Table 23: Physico-chemical characters of star gooseberry fruits

Character	Value
Fruit weight (g)	4.0-5.0
Fruit size (cm)	1.5 X 1.8
Fruit juice (%)	35-40
Pulp (%)	80
Seed weight (g/seed)	0.25
Seed size (cm)	0.5 X 0.6
Total soluble solids (^{0}Brix)	4.5 - 5
Titrable acidity (%)	2.2-2.4
pH	3.5 - 4.0
Vitamin C (mg/100 g)	35-40
Reducing sugars (%)	2.3
Total sugars (%)	3.2

It is also used in the treatment of diarrhoea, jaundice and cough. The roots and seeds are cathartic. The juice of the root bark is used as poison. It produces headache, sleepiness and death accompanied by severe abdominal pain. The root bark contains tannin (18%), saponin, gallic acid and a crystalline substance (probably lupeol). It is used in south India to a limited extent for tanning.

Soil and climatic requirements

Star gooseberry grows on all types of soils in tropical and subtropical parts of India. But, well drained deep loamy soils are ideal. It prefers moist tropical areas. But, the tree is hardy and grows successfully in degraded lands of dry subtropical regions as well.

Propagation

There are no cultivars of star gooseberry recommended for cultivation. The most common method of its propagation is through seeds. The tree can also be multiplied through budding, greenwood cuttings or air layers. The cuttings of 30 to 35 cm length having 3 to 4 nodes each should be selected from heavy-bearing trees in December-January (Kundan Kishore, 2007) for propagation.

Planting and after-care

Seedlings or cuttings after attaining 45 to 60 cm height should be transplanted in the field during monsoon season. For planting, pits of 60 x 60 x 60 cm are prepared during May-June in square system of planting. They are filled about a month later after proper exposure to sun, with a mixture of soil and farm yard manure (FYM) in 2:1 ratio. The distance between row to row and plant to plant can be kept 5 to 6 m. The tree is prolific bearer. It bears once in north India and twice in south India. As regards nutrition, application of manures and fertilizers in sufficient quantities will ensure quality production and higher yields. Since the scientific information is not available on this aspect, a dose of 5 kg FYM, 50 g nitrogen, 25 g phosphorus and 50 g potassium can be applied to a one year old plant. This dose is increased up to 6 years and thereafter 30 kg well rotten FYM, 300 g nitrogen, 150 g phosphorus and 300 g potassium should be applied every year. Full doses of manures and fertilizers should be applied in July-August with the onset of monsoon.

Irrigations should be given to young orchards for proper establishment. Young plants are susceptible to frost. Irrigations should also be given to grown up plants at 15 day intervals during summer and at 30 day intervals during winter. Mulching with organic wastes in tree basin helps in moisture conservation and control of weeds. It does not require regular training and pruning, but diseased, dead or criss-cross branches should be thinned.

Harvesting and yield

Plants start flowering 3 to 4 years after planting. In south India, trees generally yield two crops in a year, once in April-May and again in August-September. In Bengal, the trees flower mostly in hot weather and fruits can be collected in early part of the rainy season. A fully grown up tree yields up to 80 to 100 kg of fruits on an average.

References

Anonymous (1950) In : The Wealth of India-A Dictionary of Raw Materials and Industrial Products, Vol-II, pp.153-154.

Kundan Kishore, Pathak, K.A. and Rinku Bharali (2007) *Indian Hort.*, **52**(3) : 24.

Chapter 22 Paniyala

Botanical Name	:	*Flacourtia jangomas* Raeusch. syn. *F. cataphracta* Roxb.
Regional/ Vernacular Names	:	Puneala plum, paniala, talispatra, tambat, kuragayi, talisapatramu, saralu, vayangarai, chankali, garaji, kanji, talisam, vayankatha, baincha
Family	:	Moraceae

Paniyala is a shrub or erect, low-branched tree, 6 to 12 m high, has flaking bark and sharp spines on the trunk. The leaves are alternate, deciduous, pale pink when young, spirally arranged, oval-lanceolate, long-pointed, toothed, very thin, glossy on both surfaces; 5 to 10 cm long, 1.25 to 5 cm wide. Tree is dioecious, *i.e.* male and female flowers are on separate trees. They are greenish, heavily fragrant, borne in small clusters on new branchlets. The fruits are round or slightly oval, 2 to 2.5 cm long, dark-maroon to nearly black; the flesh greenish to white or amber, varying from acid to sweet, and containing 7 to 12 flat, hard, pale-yellow seeds.

Origin and distribution

The tree is native to north Bengal and east Bengal in India; commonly cultivated throughout south-east Asia, eastern Malaya, and also in the Philippines. It has been planted on limited scale in Surinam, Trinidad, Puerto Rico and southern Florida. The plant is met with in the eastern districts of Uttar Pradesh, especially Gorakhpur and adjoining areas and in Assam, Bengal, Bihar, Orissa and south India. It is often cultivated for its fruits which, under cultivation, are known to attain larger size.

Nutritive, medicinal and other uses

The fruit has a rather pleasant tart flavour. The flesh is firm, brownish green and fairly juicy. For eating out-of-hand, the fruit is rolled between the hands to reduce astringency, and is better-liked than that of other species. It is stewed as dessert, made into juice, syrup, jam, marmalade and pickles and also used in chutneys. When slightly underripe, it is used to make jelly. The acid young shoots are eaten in Indonesia.

The analysis of fruits gives 78.28% moisture, 0.03% protein, 0.39% fat, 4.86% sugar, 0.94% ash and 1.16% acidity. The fruit is fairly rich in pectin and contains 9.9% tannin on dry-weight basis.

The wood, red or scarlet, is close-grained, hard, brittle, durable and polishes well and is used for agricultural implements. The fruits are eaten to overcome biliousness, nausea and diarrhoea. The leaf decoction is taken to halt diarrhoea. Powdered, dried leaves are employed to relieve bronchitis and coughs. The leaves and young shoots, which taste like rhubarb (*Rheum rhaponticum* L.), are astringent and stomachic. The leaves and bark are applied on bleeding gums and aching teeth, and the bark infusion is gargled to alleviate hoarseness. Pulverized roots are poulticed on sores and skin eruptions and held in the mouth to soothe toothache.

Propagation

It is propagated by seeds which, however, take a long time to germinate. It can also be grown by inarching, budding or soft wood grafting/ wedge grafting on seedlings of the same species.

George *et al.* (1998) made efforts to standardize the vegetative propagation technique for paniyala. Soft wood grafting was tried in both sweet and sour paniyala types using wedge grafting. One-year old seedlings of *F. cataphracta* and *F. inermis* were used as rootstocks. Precured as well as unprecured scions of both sweet and sour types of past season growth were taken. Survival of the grafts in both open and mist conditions was observed. When compared with the sour type, the sweet rootstocks showed higher percentage of success with both sweet and sour scions. Sweet on sweet recorded 75 per cent success whereas, sour on sweet recorded 55 per cent. At the same time, sweet on sour rootstocks recorded a success of 40 per cent and sour on sour 45 per cent. As compared to precured scions, past season shoot with leaf performed better in all the four combinations of grafts. The best period of grafting was June to August in open condition. As compared to open conditions grafts kept in the mist propagation unit recorded higher percentage of success (60 to 70%) throughout the year. Though the vegetative growth of all the graft combinations was comparable, the soft wood grafts of sour type both on sour and sweet rootstocks showed early flowering and fruiting.

Planting and after-care

Seedlings or vegetatively propagated plants are planted at 6 x 6 m during July-August in square system of planting. Pits of 60 x 60 x 60 cm size are prepared a month prior to planting. Pits are filled with a mixture of top soil mixed with well rotten farm yard manure in equal quantities. Frequent irrigations are given till plants get established. There are no standard doses of manures and fertilizers recommended for this fruit crop. Tree basins are kept weed free by regular weeding.

Harvesting and yield

George *et al.* (1999) carried out studies on the flowering and fruit development of 15-year old bearing trees of paniyala in Kerala, India. The flowers produced by male trees appear as terminal or lateral cymose clusters, while those of the female trees appear in fascicles. Days taken to complete flower development were also higher (25 days) in male than in female trees (18 to 20 days). A total of 98 to 100 day period was required from flower initiation to fruit harvest. The tree flowers generally at the beginning of rains and fruits ripen from October to January. The fruits recorded the highest values for length (1.85 cm), circumference (7.36 cm), and weight (5.83 g) 80 days after fruit set. The shelf life of the fruits was low, as the scarlet colour was lost within one hour of harvest and shrivelling was observed from the second day onwards. The paniyala tree produces an average yield of 80 to 150 kg fruits per tree.

References

George, S.T., Mathew, K.L. and Mridula, K.R. (1999) *Journal of Tropical Agriculture*, 37(1/2): 17-21.

George, S.T., Mridula, K.R. and Mathew, Lyla (1998) Souvenir - Nat. Sem. on New Horizons in Production and Post Harvest Management of Tropical and Subtropical Fruits held at IARI, New Delhi on Dec. 8-9, 1998, pp. 13-14.

A Photo Gallery of Underutilized Fruit Crops

Fig. 1 : Aonla cv. NA-7 in bearing

Fig. 2 : Variability in aonla fruits

Fig. 3 : Aonla propagation by cleft grafting in polybags

Fig. 4 : Aonla popagation by patch budding in polybags

Fig. 5 : Bael cv. CISH B-1 in bearing

Fig. 6 : Bael cv. CISH B-2 in bearing

Fig. 7 : Natural variability in bael fruits

Fig. 8 : Bael propagation by cleft grafting in polybag

Fig. 9 : Carambola (Kamarakh) in fruiting

Fig.10 : Carambola propagation by cleft grafting in polybags

Fig. 11 : Custard apple in fruiting

Fig. 12 : Custard apple propagation by cleft grafting in polybags

Fig. 13 : Wood apple in fruiting

Fig. 14 : Variability in wood apple fruits

Fig. 16 : Tamarind in fruiting

Fig. 15 : Wood apple propagation by cleft grafting in polybags

Fig. 17 : Jackfruit in fruiting

Fig. 18 : Lasora in fruiting

Fig. 19 : Lasora propagation by cleft grafting in polybags

Fig. 20 : Lasora propagation by patch budding in polybags

Fig. 21 : Loquat cv. Tanaka in bearing

Fig. 22 : Loquat cv. Improved Golden Yellow in bearing

Fig. 23 : Loquat propagation by cleft grafting in polybag

Fig. 24 : Barbados cherry in fruiting

Fig. 25 : Branches of chironji tree

Fig. 26 : Chironji fruits

Fig. 27 : Jamun tree in fruiting

Fig. 28 : Jamun–a fruited branch

Fig. 29 : Jamun propagation by cleft grafting in polybags

Fig. 30 : Phalsa in fruiting

Fig. 31 : Fig in fruiting

Fig. 32 : Mulberry- a fruited branch

Fig. 33 : CISH Mulberry-2

Fig. 34 : CISH Mulberry-8

Fig. 35 : Karonda in fruiting

Fig. 36 : Cape gooseberry in fruiting

Fig. 37 : Cape gooseberry fruits

Fig. 38 : Packing of cape gooseberry fruits for sale

Fig 40 : Promising cultivar of barhal

Fig. 39 : Khirni in fruiting

Fig. 41 : Variability in barhal fruits

Fig. 42 : Amra in fruiting

Fig. 43 : Star gooseberry in fruiting

Fig. 44 : Paniyala in fruiting

Index